Illustrated Guide to Fortrose and vicinity, with an appendix on the antiquities of the Black Isle and a map.

John Angus Beaton

MAP OF THE

BLACK ISLE

ROSS-SHIRE.

SPECIALLY PREPARED TO ILLUSTRATE
GUIDE TO THE DISTRICT, BY

A. J. BEATON, C.E., F.S.A. Scot., M.S.G.S.

1885.

English Miles

FORTROSE CATHEDRAL. (SOUTH SIDE)

ILLUSTRATED GUIDE
TO
FORTROSE AND VICINITY,

WITH AN APPENDIX ON THE

Antiquities of the Black Isle.

AND A MAP.

BY
ANGUS J. BEATON, F.S.A.Scot.

INVERNESS:
WILLIAM MACKAY, 27 HIGH STREET.
1885.

TO

JAMES FLETCHER, Esq., D.L., J.P.,

OF ROSEHAUGH, LETHAM GRANGE, ETC.,

IN GRATEFUL RECOGNITION OF

THE INTEREST

HE HAS

ALWAYS TAKEN IN THE

SOCIAL, MORAL, AND INTELLECTUAL

WELFARE OF THIS COMMUNITY, AND FOR HIS MANY

BENEVOLENT DEEDS AND CHARITABLE ACTS

SO FREQUENTLY PERFORMED IN

THE BLACK ISLE,

THIS LITTLE VOLUME

IS RESPECTFULLY DEDICATED BY

THE AUTHOR.

PREFACE.

IT is not without hesitation that I launch this little volume on the stream of public opinion; being fully alive to its many deficiencies, yet relying on the generosity of the public, I venture to solicit a lenient criticism of a production resulting from a desire to bring into prominence a locality so rich and interesting from an antiquarian point of view, and so healthful and salubrious from a hygienic stand-point. A work of this nature must consist more or less of compilations; but, while making copious extracts from valuable resources, I have endeavoured as much as possible to avoid the stereotyped phraseology of *bona fide* " Guides," by intermingling a few current traditions of the district, and offering matter that may prove as interesting and instructive to the every-day resident as to the flying visitor. Also, in preserving some traditional lore, and presenting to the public an epitomised history which many would find considerable difficulty in obtaining, owing to the scarcity of old papers and statistical accounts from which they are copied.

I have arranged a series of tours which can be accomplished with ease from Fortrose as a centre, noting all places of interest, and added as an appendix a brief historical sketch of Cromarty, as well as a classified arrangement of antiquities and places of interest in all the remaining parishes of the peninsula ; these are but the simple aggregation of facts and ideas that have occurred to me in my rambles in the Black Isle, but which I trust may form the nucleus of more erudite matter, which I hope at some future period to produce, and any information or suggestion for improvement will be gratefully received.

In compiling this little work, I have to acknowledge the assistance and valuable information so readily supplied by the following gentlemen, viz. :—Mr JOHN HENDERSON, Town Clerk, Fortrose, in furnishing copies of old papers ; Mr W. S. GEDDIE, Fortrose ; Messrs JAMES FRASER, M.Inst.C.E., and JAMES ROSS, F.E.I.S., Inverness.

The map, which is revised and reduced from the Ordnance Survey, is specially prepared for this Guide by BARTHOLOMEW, Edinburgh, and the views are from photographs expressly taken to illustrate the work by Mr W. D. CLARKE, Inverness, and photo-lithographed by Messrs AVERY & Co., Aberdeen.

CONTENTS.

LIST OF ILLUSTRATIONS.

ERRATA.

	Page
For " judicial," read " judicious " .. (11th line) ...	31
Runic Cross of Rosemarkie should read " Photographed from the stone," instead of " from a drawing "	49

THE BLACK ISLE.

THE BLACK ISLE is a peninsula, bounded on the north by the Cromarty Firth, and on the south by the Moray, Inverness, and Beauly Firths. How it acquired the epithet "Black" is doubtful. One or two of the most probable solutions I take the liberty of advancing. Sir John Sinclair ascribes it to the dark and heathy common called the Mael Bhuie, or Millbuie, signifying the yellow ridge or moor, which traverses the whole length, and forms, as it were, the backbone of the peninsula, which in remote ages presented a dull and doleful appearance. The original appellation was Edderdail, or the land between the two arms of the sea. This might in course of time become corrupted to Ellandhu; the former, however, seems most probable. The more modern name was Ardmeanach, the height in the middle, but called by some Ardmanach, the land or territory of the monks. Either of these names are very appropriate—the peninsula being formed like a house ridge (anticlinal), and monks once possessing considerable lands at "Fortrosse," "Rosemarkyn," and "Beulie."

The lordship of Ardmanach, or the Black Isle proper, comprises part of four counties, viz., Ross, Cromarty, Inverness, and Nairn, and contains eight parishes. We find it chronicled that "in 1179 King William the Lion, his brother David and nobles, visited Ross-shire to compose some disorder in that distant quarter, and built Red Castle in the Ardmanach." *

* Redcastle, in the parish of Killearnan, built near the shores of the Beauly Firth, is the seat of the Hon. H. J. Baillie, and the oldest inhabited house in Scotland. See Killearnan Parish.

The Black Isle has been the scene of many a wild tale of superstition. Its legendary personages, including the still dreaded witches of Ferrintosh (possibly the conjuring effects of its far-famed whisky); the giants of Craigiehow cave, who are at any moment, on the sounding of the trumpet, ready to sally forth and destroy the world; Ipack of Ord Hill, whose midnight orgies were the terror of the neighbouring lands, as she scudded along Munlochy Bay and the Firth in a "lippy measure," and broomstick for a sail; and last, but not least, Coinneach Odhar, the Brahan Seer, who, according to story, met his untimely fate at Chanonry Point. It is possible that in these superstitions we may trace the origin of the epithet "Black Isle"; or more probably the prevalence of ancient remains in the district accounts for the varieties of superstitions.

It is declared that the Black Isle can bear comparison with any district in Scotland, not only for its enterprise in agriculture, its lovely scenery, and salubrious climate, but for its importance in the olden times, and in no part are these more demonstratively exhibited than in the lovely burgh of

FORTROSE AND ITS VICINITY.

The ancient town of Fortrose, once the principal seat of ecclesiastics and literature in the North, is picturesquely situated on the shores of the Moray Firth, which from its romantic position possesses many charms for the tourist and pleasure seeker, while its salubrious climate should attract those desirous of imparting the ruddy glow of health to the pallid cheek.

Before venturing to further describe Fortrose, we will imagine the reader on board the tidy little steamer "Rosehaugh," steaming from the Thornbush Harbour, Inverness; and will notice anything of interest as we glide down the smooth waters of the Firth. On emerging from the River Ness, and looking down the estuary, the idea conveyed is that we are on the bosom of some inland lake, because Chanonry Point and the Ardersier Point appear united, and enclosing the basin called the Inverness Firth. The high and precipitous hill opposite the mouth of the river is called the Ord Hill of Kessock, consisting of old red

FORTROSE, LOOKING TOWARDS INVERNESS.

sandstone conglomerate, a formation which traverses the Black Isle to the Souters of Cromarty, and extending on the Inverness-shire side up the north side of the Great Glen as far as Dunain Hill. A glance at the map cannot fail to attract the attention to the remarkable line the north coast forms, being almost a straight line from Ord Hill to Caithness-shire, and a continuation of the line of the great glen from Fort-William. Ord Hill, 633 ft. high, has a vitrified fort, or rather a line or wall of vitrified matter, extending along the edges of the cliffs, and inclosing an area of two hundred yards by sixty yards; on the west side, where the rocks are less precipitous, the remains of a strong barrier of loose stones are seen. Craig-Phadrig, the conical hill on the opposite side of the Ferry, also a very fine example of a vitrified fort, has a tendency to be oval in form, its greatest length being 66 ft. by 100 ft. broad, and the vitrified walls from 6 to 8 feet thick on average. It is an excellent specimen, and would well repay a visit to those interested in this curious class of antiquities.

Continuing our journey down the Firth, we notice the small village at the base of Ord Hill, called Kilmuir, with the mansion house of Drynie on the terrace above. The ruined church further east is that from which the village takes its name: the church was dedicated to St Mary, hence the name *Cille Mhuire* (Gaelic). The parish here takes its name from this church, but strange, on its union with the adjoining parish of Suddie, the name was changed to Knockbain (Whitehill), a transformation which speaks volumes for those would-be improvers of our nomenclature and their love of the antique!

Beyond the ridge which rises behind the church of Kilmuir is Loch Lundie, said to be of unfathomable depth, and the abode of a water sprite; on the north side of the Loch is a rock called " Craig-an-Caisteill," on which is a small earth fort, of an oval form, measuring inside the walls from east to west 54 feet, and 43 feet north to south at the widest part. The wall is 4 feet thick all round, except at the west side; where the natural defences are less protective it is 7 feet thick and somewhat higher.

We now arrive at Munlochy Bay entrance, which is guarded on either side with a headland or abruptly terminating rock, conveying the impression that at some remote period those two bluffs were united and had been rent asunder by enormous pressure. A casual observation of the Munlochy and Rosehaugh valleys tends to confirm this supposition : an identical example may be seen at the Souters of Cromarty.

The rock on the west side is called Craigiehow (Creag-a-chow), and has a pretty extensive cave. Of the many wild superstitious legends connected with the cave, these are among them : On one occasion a marriage party, exhilarated with the joyful excitement with which such an event imbues its participators, determined to explore the reported mysteries of the cave ; headed by a piper, they pushed on through the damp cavern, until, feeling desirous of returning, they found the open way they passed through closed firm as the solid rock ; forward the way was clear, but each successive chamber they entered and left had its ponderous door open for their entrance and immediately close after them. Hearing the rushing of waters, one of the party, with wonderful presence of mind, wrote a narrative of their fatal adventure, after taking a parting "quaich of usequebey," * and having deposited the epistle in the empty bottle, pushed ahead, vainly hoping for some means of escape. Next morning the bagpipes and the bottle were found floating on Loch Lundie, proving that the cave terminated in the Loch—the wonderful bottle solving mysteries hitherto unknown and sealing the doom of the unfortunate explorers.

Another superstition is that the Fingalian hunters are imprisoned here, and that when the silver trumpet which lies on the stone table of their prison is blown, they rush forth to destroy the world. A dripping well at the mouth of the cave is resorted to till this day by persons to get cured of deafness. The conical hill on the east side of the Bay is called Ormond, or Lady Hill, and has the ruins of a most interesting castle on its eastern shoulder. The castle is minutely described in a succeeding chapter. The highest points of Ormond Hill and Craigiehow are 390 feet and 400 feet respectively.

We now round the Bay of Avoch, and notice the fishing

---

* At all Scotch marriages and funerals it was customary to carry whisky along with the party, and treat every one they met on the way with a dram, a custom we are glad to see dying away with the superstitious generation with whom it originated.

10

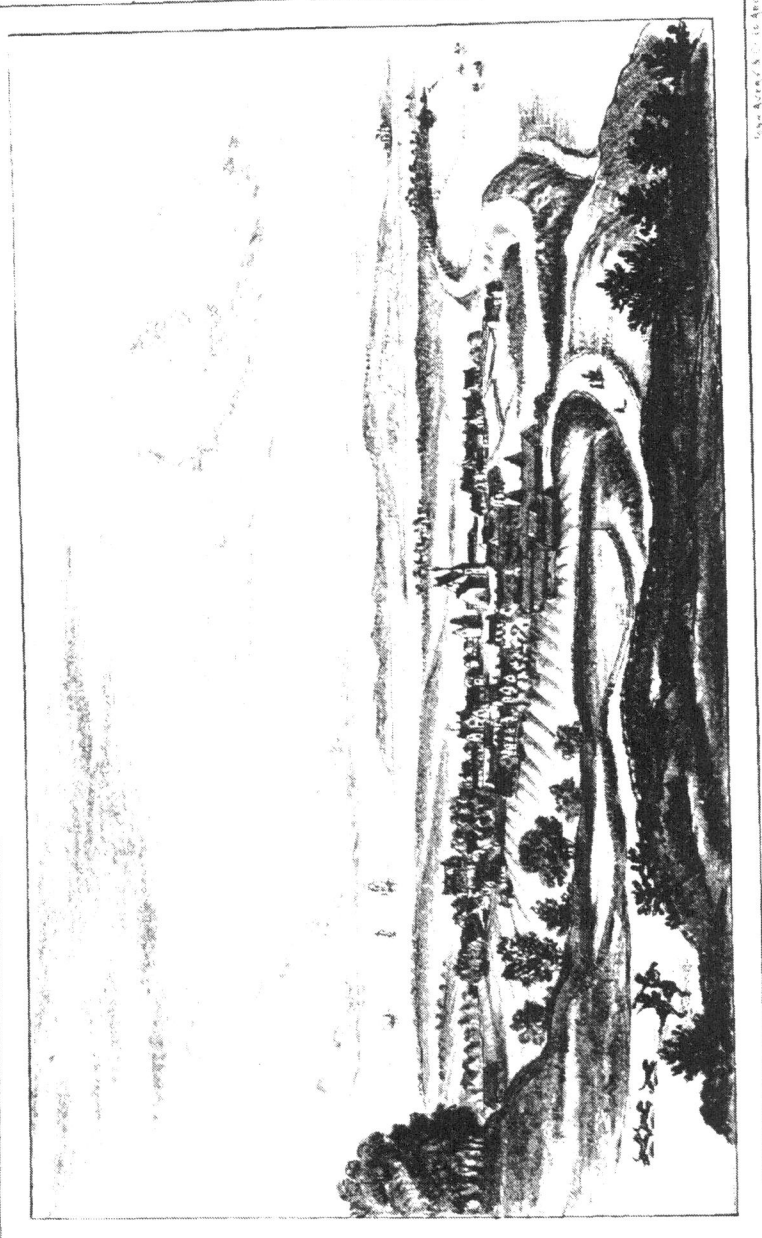

CHANONRIA CIVITATIS ROSSIÆ. THE CHANNERY TOWN OF ROSS, 1693.

village with the wooded vale of Rosehaugh stretching up behind (this part is referred to further on) As we steam along inshore, between Avoch and Fortrose, we have a fine view of the Craigwood and cliffs. These cliffs are an upheaval of quartzite. We hope soon to see the Burgh acquire a right to traverse the wood, with walks for public recreation. We now arrive at the ancient town, and, if the tide should happen to be low, we land by the substantial jetty recently erected at a cost of between £6000 and £8000. It is a valuable acquisition to the burgh, and forms a pleasant, cool promenade in warm weather.

Having arrived at Fortrose, we will proceed to give a brief sketch of its history, noting places of interest within its bounds, and epitomising descriptions of short and interesting excursions, which may easily be made from this centre.

FORTROSE.

Ancient History.

F ortrose, historic and ancient,
O ld memories cling to thee ;
R est in thy peaceful beauty.
T hou lovely sea-girt Chanonry.
R ichly dowered by nature,
O n every side thou'rt blest ;
S ol's last warm rays fall on thee,
E 'en as he sinks to rest '
—*Mary Stewart, Inverness.*

THE ancient town of Fortrose, once the principal seat of ecclesiastics and literature in the north, is picturesquely situated on the shores of the Moray Firth ; from its romantic situation it possesses many charms for the tourist and pleasure-seeker, while its salubrious climate and excellent sea-bathing beach may classify it first among our northern watering-places and health resorts.

The curvature of the coast line, with its fine sandy beach, forms two remarkably fine bays, the Chanonry promontory acting as a natural breakwater, thus always affording one smooth and calm bay, blow the wind from any point. Either of the bays is admirably adapted for sea-bathing or yachting, but we would suggest the advisability of having several bathing machines both at Rosemarkie and Fortrose. At most watering-places, many visitors attend for pleasure, and even the invalid requires occasionally some amusement to break the dull monotony of sea-side life; if, therefore, the Commissioners would awake from their apathy, and engage a good band to play on the pier, charging a nominal fee for admission, they would confer a boon on the visitors and a benefit on their own constituents.* Another suggestion : with comparative little outlay a fine esplanade could be formed from the present harbour, westward to the Craigburn bridge, and having connection through the Craigwood, where pretty paths might be cheaply constructed. These may seem extravagant suggestions to frugal and canny Scots, yet we have noticed more important improvements remuneratively executed in places not possessing half the natural attractions of Fortrose. There are many pretty walks around the burgh, and pleasant tours may easily be made into the country to view objects of interest and antiquity, but these are referred to under a classified list of excursions.

The ancient name of the Burgh was Chanonrie, but in 1444 it was united to the Burgh of Rosemarkie by a charter from King James II., under the common name Fortross, now softened into Fortrose ; which charter was ratified by King James VI. in 1592, and confirmed in a still more ample form by the said monarch in 1612. The following is a copy of the charter granted by King James II., dated 18th June 1455 :—

"James, by the grace of God King of Scots, to all his good subjects to whom these presents shall come, Health. Know ye, Because we, on the motion of our divine friendship, and for the singular favour, zeal, and

* Since the above has gone to press we are glad to learn that a good brass band has been organised.

FORTROSE HARBOUR AND LANDING PIER.

love which we bear towards the venerable Father in Christ, Thomas, Bishop of Ross, and for the augmentation of his ecclesiastical liberty, have given, and by the tenor of these presents, for us and our heirs and successors, heritably and perpetually give to the said venerable Father in Christ, Thomas, Bishop of Ross, and his successors of the said church bishops that they (ipe), holding and inhabiting the village of Fortrose, now called Canonia, in which is situated the said Church of Ross, may have, hold, and possess in times to come, perpetually, that village as the free burgh of Fortrose, and as freely with all and singular, the privileges, liberties, customs as the said Burgh of Rosemarky, inferior with its liberties, commodities, and easements is more freely seized. Which village of Fortrose we have annexed, incorporated, and united, and by the tenor of the present writing, for ever annex, incorporate, and unite to the said burgh of Rosemarky, and moreover we will, and for us and our successors, Kings of Scotland, grant by these presents, that those holding and inhabiting the said village of Fortrose and Rosemarky be free and exempt and quit of all *forensico serviio auscibus prisus, ariagus taxatiorubus exacticnibus oneribus, aut servitius sicularibus quibuscunqre.* As in the ancient charteris and *tris* of the late serene prince Alexander King of Scots, and of our other predecessors, kings of Scotland, concerning the said privileges of the said Burgh of Rosemarky granted is more fully contained. And that those inhabiting the said Burgh of Fortrose may enjoy and possess all the liberties and privileges pertaining to the said Burgh of Rosemarky in all future times, as freely as those inhabiting the Burgh of Rosemarky at any time past more freely enjoyed and possessed its limits. Wherefore we Command all and every one of our lieges and subjects, strictly enjoining them that no one presume in any way to attempt in contradiction of our said grant, under all the pains which can be incurred against our Royal Majesty. Given under our great Seal at Edinburgh, the eighteenth day of the month of June, in the year of our Lord one thousand four hundred and fifty-five, and of our reign the nineteenth."

The following extracts from an Act of Parliament (May 1661), confirming the union of Fortrose with Rosemarkie, are most interesting and amusing. It will be seen that at this date the ancient burgh of Rosemarkie was much decayed and almost entirely deserted, while Fortrose was increasing " more and more."—

" At Edinburgh, the twentie day of May, and yeir of God One Thousand Six Hundred and Three Score yeires, Our Soveraigne Lord and Estates of Parliament, Considering that His Majestie's single dearest giude King James the Sixt, of ever blessed memorie, by his Charter under His Majestie's great Seal of the dait at Edinburgh, the sixt day of August, One thousand five Hundred fourscore and ten yeires . . . . to ane uther infeftment, Grantit be His Highnes' most noble progenitors, also under the great Seal, Erecting and making the Toune and Lands of Fortrose, of old called the Chanonrie of Ross, in ane frie burgh for the onerous causes and gude considerations moveing His Hieness thereto, not only for the weill and utilitie of the said burgh of Fortrose and inhabitants thereof, but also for the profite and commoditie of whomsoever His Leidges neir and about the same, and continuallie repairing hither." It goes on granting power to the inhabitants to choose and elect " Provoist, Bailyies, Councillors, Decons of crafts," &c., with power of buying and selling of " merchandise and giudes suck-lyke,". . "and possess it and enjoy the haill priveleges, immunities, and perogatives, fullie as anie uther frie burgh within the realme . . . or anie uther frie burgh are in use to doe posses, and to have Provoist, Bailyies, Counsell, and uther officers requisite within the said Burgh for goverment thereof and administration of justice . . . with frie power, libertie of buying and selling of wyne, wax, cloth, linnen and wool, salt, piekle, tarr, honey,   and keip therein baxteries and slayers of fish and flesh, and all uther artificeris belonging to a frie burgh . . . . . . and also have granted and disponed to the said   Provoist, Baillies, connors thereof, and their successors two meiklie mercat days, the one to be on Pardon Munday, and the otter on the Sabbath day."

I wonder what the orthodoxy of the nineteenth century would think of such a proclamation as holding a "muckle market" on the Sabbath-day. These markets were eventually ordered to be held on Tuesday and Friday for ever. In addition to these markets, two free fairs were to be held yearly, "for the buying and selling of all kynde of giudes and merchandese, and the said mercates to indure for the space of three dayes after the proclaiming thereof, as in former tymes was accustomed."

Politicians will read the following letter with admiration at the tact employed to redeem the favours of a formerly rejected candidate. It is addressed to Lord Fortrose by the Magistrates and Town Council of Fortrose, but, unfortunately, no date is appended. Possibly it was written shortly after the annexation of the two burghs. The letter begins with a plea for their being deceived at the last election, and winds up with a deplorable account of the extreme poverty of the Burgh, and the load of taxes they groan under. From its unique style, I give the letter in extenso :—

"My Lord, we are honoured wt. yours from Edr., of date 27th of

OLD HAND BELL OF TOWN OF FORTROSE.

July last, intimating to us your intention to offer your service to the district of Burrows, to serve for them in the ensuing parliat. We do acknowledge wt regret that it was not our good fortune to have voted for your Lordship at the last General Elections. We were then fed with false musick, hoodwinked and really not ourselves, but since that occasion such occurrencies has happened as makes this community in general and the Magistrates and Council of the same in particular, of opinion that the district of Burrows, in which we are classed, will be highly honoured in having your Lordship their Representative in Parliament, and what confirms us in this opinion the more, is that after looking back on what happened at or since those General Elections we have no manner of reason to complain of the least impression on your Lordship's mind to the prejudice of any individual person in this corporation. The asserting our own liberty, and being masters of ourselves, we still look to be the duty of all burgesses, yet we cannot but wt gratefull resentment thank your Lordship for taking a particular notice of it, and thereby approving our actions. And in this situation your application for our favour and friendship is very acceptable, and with pleasure we can assure you that as we think no person has as yet deserved better of your country than your Lordship, so none will byass our interest or countenance in your Lordship's prejudice at the ensuing Elections. This day we have the pleasure of Achnagairn's company, with whom we drunk your Lordship's health and your good success in the ensuing Elections, who has communicated and left with us Provost Drummon's letter to your Lordship, with the Act of the Convention of Burrows in our favour. We own our obligation to your Lordship for the least notice and concern you take for the ease of this poor Burrow, and will always entertain a grateful resentment of it. Meantime, we beg leave to observe that the Provost's letter mentions only the giving us ease of a penny in the Tax Roll, and if the Committee can give us no furder ease, it will be of no avail, and we will have still the same cause of complaint, and groan under our burden as formerly. We are now sevin pence in the Tax Roll, which is a burden that this poor place is in no condition to bear. We at least expected the dimunition of a groat or three pence, and in caise the Committee are not in power to affoard that ease, after a due report, the design of the Act will be entirely ineffectual to our purpose. Our stent is the double of some oyr burrows who are in better condition to bear seventeen pence than wee to bear seven. Your Lordship will please excuse, &c."

The following is a portion of a minute of

"A Council Meeting holden at Fortrose, the twenty-second day of September 1709, by John Dallas [of Bannans] one of the Baylies (Hugh Baillie, one of his colleagues, being necessarily absent at Ednr. for the time *republicae causa*, and the other Baylie being elsewhere).

"The said day the said John Dallas produced a disposition granted be George, Earle of Cromarty, wt. consent of Sir James Mackenzie, his son, in favour of the Magistrats, to the behoof of the poor ones of this place, date 6 July 1709, of and upon the fyne, pecks and ane half land extending to twenty-two bolls, and ane half rent (altour few) lyeing inabout Channonry and Rosemarky, Dyspond to the Earle forsd. for the sum of one thousand six hundred and eighty-seven pounds ten shillings Scots money advanced to the Disponer as price yrof, compting the same at twelve hundred pounds each chalder, which was unproven out of the original stock by the Magistrats, of one thousand merks scots, piously doted and mortyfied to the sd. poor be the deceast Alexander Mackenzie, sometime of Coulle, great-grandfather to Sir John Mackenzie, now of Coulle."

The minute goes on describing the various purchases, &c., of the land, which, at that time, seemed pretty much mixed up. Eventually, we find an obligation procured from the Laird of Cromarty for the progress of the purchases, which winds up with an expression of public thanks to Bannans for his pains and assiduous care in purging and settling matters.

Fortrose in 1622 was spoken of as a town "flourishing in the arts and sciences, being at that period the seat of divinity, law, and physic in this corner of the kingdom."

Most of its ancient glory is now faded; its castle is razed to the ground; the bishop's palace and the residences of other dignitaries share the same fate; even its once grand cathedral presents but a fragment to convey to the inquirer a glimpse of its past grandeur; yet with all its modern improvements one feels the sanctity of its antique atmosphere while walking round the cathedral square, and very little imagination will conjure up the procession of surpliced priests and choristers chanting some weird anthem through the one remaining aisle, and it may not be too imaginative to suppose that at some future age it may be restored, if not to its once extensive area, at least to a comfortable place of worship.

Though much of its antiquarian interest is now destroyed, still all its natural beauties and attractions remain as they did seven centuries ago, when a few lone monks selected this as the most pleasant situation in the northern regions, and settled down on its placid shores, inhaled the balmy atmosphere of its secluded walks, and laid the foundations of an institution which may have done good work in those dark and superstitious

INTERIOR OF FORTROSE CATHEDRAL (LOOKING WEST)

ages, withstanding every conflict, until the advent of Cromwell's blood-thirsty Commonwealth hurled it to oblivion.

The Cathedral.

The Cathedral Church, dedicated to Saint Peter and Saint Boniface, stood at the Canonry (now Fortrose) [the Gaelic for Fortrose is Cannanich], about a mile west from the parish church [at Rosemarkie], When entire, it consisted of choir and nave, with aisles, eastern Lady chapel, western tower and chapter-house, at the north-east end ; its remains consist of the " south aisle to chancel and nave, and the detached chapter-house," all in the middle-pointed style. The Seal of the Chapter, now used as that of the Burgh, bears the figures of Saint Peter and Saint Boniface, and the inscription, SIGILLVM SANCTORVM PETRIE ET BONIFACII DE ROSMARKIN.

A large old bell, now hung in the modern spire of the Cathedral, bears the name of Thomas Tulloch, bishop of Ross, the date 1460, and an inscription intimating its dedication to the Virgin Mary and Saint Boniface. The plate opposite represents the old town hand-bell, bearing the following inscription : PRE-ISIMUS NON AMISIMUS, 1630.—CHANRIE DE ROS, (with bishop's mitre between) R. H.

"The Bishoprick of Ross was founded by King David I., between the years 1124 and 1128, at which period, Macbeth, its first Bishop *(Rosemarkensis episcopus)*, appears on record. The bishop's seat was perhaps originally at Rosemarkie, from which the bishoprick continued to be occasionally named down to the middle of the thirteenth century, when the original name seems to have finally given place to that of bishoprick of Ross. Among the items of the King's revenue accounted for in 1263 by Laurence le Grant, Sheriff of Invernes, was the king's silver *(finis)* paid by the bishop of Ross, amounting for that year to 10 marks, and the profits *(lucra)* of the justiciar in Ros, amounting to £4 10s., exclusive of the Bishop's tithe, which was 10s. In 1329 the Abbot of Dunfermline, depository of the money ordained *pro pace* (for fulfilling the stipulations of the

Treaty of Northampton, 1328) accounted to the King's ex-
chequer for the sum of £71 11s. 1d., received from the bishoprick of Ross. From that period till near the end of the fifteenth
century there seems to be almost no recorded notice of the
bishoprick, except in the case of the successive rulers. It is
supposed that all the records were carried by Bishop Leslie to
Paris or Rome, but it is possible they may be hidden about the
cathedral. John Fraser is alleged to have been appointed bishop
in 1485 . . . . . In 1507 (1st May) that king granted
for a year to Sir Robert Fresale, Dean of Ross, and Alexander
Fresale, James Makysoun, the executors of the deceased John,
Bishop of Ross, the temporality of the lands and possessions
of the bishoprick, with power to sublet and hold courts, with
all other liberties.

"In 1584 (4th February) King James VI. confirmed a charter by Henry Kincaid, rector of the parish church of Lymnolair,
and canon of the Cathedral Church of Ross, granting to Master
John Robertsonn, treasurer, and his wife Elizabeth Baillie, and
their heirs, with remainder to John's heirs whomsoever, the
croft called Lymnolair, lying with the bounds of the Canoury."

"In the Cathedral Church was a number of chaplainaries;
the chaplain of which held some lands and revenues in common.
. . . . . In 1580 King James the sixth granted in heri-
tage to Robert Grahame the three crofts of land called the land
of the chaplains of the Cathedral Church of Ross, and belonging
to the chaplains and stallers there founded ; two of which crofts
without houses lay between the common lands of Roismerkie on
the north and south—the lands of the bishoprick let in feuferme
to Colin Mackenzie of Kintaill, the common road between the
Canonry of Ross and Roismerkie, and the common lands of
Rosemarkie on the north (east?)—and the green path*(*transitum
vividem*) between the Canonry and the fishertown of the same
on the west ; and the third croft with houses had a piece of land
called the Bischopis Sched, then let to the same Colin in feuferme on the east—the public street of the Canonry on the

---

* The green path evidently refers to the beach extending between
Avoch (the fishertown) and Fortrose ; possibly before the road was constructed, the "links" might have extended all along, as they now do at
the ness or point.

DIOCESAN SEAL OF FORTROSE OR CHANNONRY.

south—the path between the canonry and 'lie Plotcok' on the west—and the community (common lands and houses) of Plotcok on the north; with reservation of the usefruct and life-rent to the existing chaplains and stallers; the grantee paying yearly a silver penny at the cathedral on the feast of Pentecost to the King, if asked, and the sum of £20 Scots at the usual terms to the master of the grammar school of the town of the Canonry of Ross."—the Kintail mentioned here as feuing the lands being undoubtedly the "Seaforth" who drove the Munroes of Foulis, the ancient custodiers, out of the castle, and held undisputed sway for many years—

> " For who in the land of the Saxon or Gael
> Could match with Mackenzie, High Chief of Kintail !"

Not a vestige now remains of the once formidable structure, and the then powerful Mackenzies declined so rapidly that even Sir Walter Scott, in his day, wrote that

> " Of the line of Mackenneth remains not a male
> To bear the proud name of the Chief of Kintail."

In the Register of the Privy Council, we find recorded that Rory Mackenzie, brother to Colin of Kintail, was put to the horn in 1673, at the instance of Mr George Monro, who had been appointed Chancellor of Ross, and who complained that " Rory, brother to Colin Mackenzie of Kintail, having continual residence in the steeple of the Chanonry of Ross, which he caused big not only to oppress the country with masterful reef, sorning, and daily oppression, but also for suppressing of the Word of God, which was always preached in the said Kirk before his entry thereto—but is now become a filthy sty and den of thieves—has masterfully and violently, with a great force of oppressors, come to the tenants indebted in payment to the said Mr George, and reft them in all and haill the fruits of his benefice." Mr George complains further that through fear of his life the oppressor compels him to refrain from discharging the duties of the vocation to which God had called him. On the 26th of March, 1575, Colin, Earl of Argyll, and Robert Monro of Foulis, became sureties that Rory shall return to Regent

Morton a bond of Walter Urquhart, Sheriff of Cromarty, John Grant of Freuchy, and Hucheon Rose, Baron of Kilravock, obliging them to enter the said Rory before the Council when required, and that he shall in the meantime keep good rule in the country.

"During the 'covenanting times,' Ormond Castle was occupied by Sir George Mackenzie, called the "Bluidy Mackenzie." It was chiefly owing to the close proximity of Ormond Castle to Fortrose, and the fear of the "Bluidy Mackenzie," that Leslie, Bishop of Ross, maintained his position in Ross-shire when all the rest of Scotland was in such excitement; Leslie, Bishop of Ross, had a great hand in framing the new prayer-book, which caused such a noise in Edinburgh when "Jenny Geddes" flung her " creepy " at the Dean's head.

Leslie, backed by the "Bluidy Mackenzie" and his troops at Ormond, thought that he at least would have the new Liturgy introduced to Fortrose, but the Bishop reckoned without his host ; for if the good people of Fortrose were afraid of the "Bluidy Mackenzie" and Bishop Leslie, the boys attending the Grammar School made up for their seniors' temerity, and on the quiet Sabbath morning forced their way into the Cathedral, gathered all the new prayer-books, rushed down with them to Chanonry Point with the intention of having a bonfire : the wind, however, being very high, extinguished the torch they carried ; they had, therefore, to content themselves by tearing the books, and throwing them into the sea. Bishop Leslie, however, had the good sense to go on with the service as if nothing unusual had transpired, but no sooner was the service over (which, by the way, was one of the shortest ever dispensed), than he made across Fort-George Ferry, and rested not till he took refuge with his brother bishop in Spinie, near Elgin. Both Bishops, burning with anger, made their way to the Court of Charles, but too late to have matters remedied.

"In the year 1226, a controversy between Robert, Bishop of Ross, and John Byseth, about the patronage of the Church of Kyntalargy, was settled as follows : The Bishop, with the consent of the Chapter of Rosemarkyn and his other clergy of Ros, quit-claimed to John Byseth and his heirs, for their homage, his right of patronage, if any ; and John Byseth and his heirs quit-

THE CROSS, AND HIGH STREET, FORTROSE.

claimed to the Bishop whatever right they had to the kirkland of the said church ; and John Byseth, besides, for the purpose of settling the controversy, and as an atonement for his own sins *(pro redemptione peccatorum suorum)*, contributed 15 marks of silver to the fabrick of the Church of Saint Peter of Rosmarkyn and a stone of wax yearly from himself and his heirs, to light upon the altar of that Church. . . . . . . In 1227, on the settlement of a dispute between the bishops of Moray and Ross, about the Church of Kyntalargy and Ardrosser, the Bishop of Ross gave up the stone of wax thus acquired, for the use of the Cathedral Church of Elgyn. There is a legend that in sending the plans of the Elgin and Fortrose cathedrals from Rome, the one intended for Elgin was sent to Fortrose, and the Fortrose plan forwarded to Elgin ; there may, therefore, be some connection between this and the superstitious tradition current that the fairies exchanged the buildings when finished. In 1388 Sir Andrew de Moravia, lord of Bothwell and Avoch, died in Ross, and was buried in the 'Kyrk Cathedral of Rosmarkyne.' "

The following description of the Cathedral is from Mr Heale's "Ecclesiastical Notes" of 1848, and gives an elaborate and thoroughly professional description :—

"The Cathedral formerly consisted of choir and nave, with aisles to each, eastern lady chapel, western tower and chapter-house at the north-east end. What remains consist merely of the south aisle to chancel and nave, and the detached chapter house. The style is the purest and most elaborate middle pointed ; the material, red sandstone [possibly from Arkendeith or Suddie Quarry], near Munlochy, gave depth and freedom to the chisel ; and the whole church, though probably not 120 feet long from east to west, must have been an architectural gem of the very first description.

"The exquisite beauty of the mouldings, after so many years exposure to the air, is wonderful, and shows that in whatever other respect these remote parts of Scotland were barbarous in ecclesiology, at least they were on a par with any other branch of the Mediæval Church.

"The east window, fragments of the tracery of which hang from the archivault, must have been magnificent, and consisted of five lights ; it is wide in proportion to its height, and must have afforded great scope for showing up the altar beneath : on the outside of the gable there are two lancets, the lower one much longer than the other. The whole effect is extremely satisfactory. I know not, indeed, where one could look for a better model for a small collegiate church, and such as might suit the needs of our communion at this moment. There are two windows in the south side, of the same elaborate and beautiful description, but consisting of four lights. The Piscina remains, and the mouldings are truly the work of a master. The south aisle was separated from the chancel by two middle-pointed arches, now walled up [these have been opened since this was written] but not so much injured as to destroy their extreme loveliness. In the first of these arches is a canopied tomb for the foundress, a countess of Ross, the date of which is probably 1330. Very possibly her lord might be interred in a similar position in the north side of the choir. This must have been one of the most beautiful monuments I ever saw. Between the foot and the eastermost pier a credence is inserted, sloping up with a stone lean-to against the passage wall. In the second arch is a pure third-pointed high tomb and canopy, with the effigy of a bishop, by tradition the second Bishop of the See : a thing manifestly impossible unless the monument were erected long after the decease of the person commemorated. The chancel arch is modern. The nave consists of four bays, and much resembles the chancel in its details ; the fourth is, however, blocked off for the burying place of the Mackenzies of Seaforth. In the second arch is another third-pointed monument ; on the south side the first window is injured ; the second resembles those in the chancel arch ; the third is high up and mutilated ; the fourth is a plain lancet. The west front is remarkably simple, and contains nothing but a small two-light middle-pointed window, without foliation. The Rood turret still exists, and is very elegant, though of somewhat singular composition. It stands at the junction of the south aisle of nave and chancel, and acts as a buttress. Square at the base, it is bevelled into a semi-hexagonal superstructure, and has elegant two-light windows on alternate sides. The top is modern. The chapter-house, as at Glasgow, consists of two stages, a crypt and a chapter-house, properly speaking. The crypt still remains, and the upper part, which was rebuilt, is now a court-room and used as the Town Council meeting-room."

The ground plan inserted will give an idea of the original extent. The portions still standing are shown in *black*, while the trace of foundations are indicated by *dotted* lines From this will be seen that the total original length was about 192 ft., and the width of nave, choir, and lady chapel, about 27 ft.— both inside measurements. The remaining portions, viz. the side aisle of nave, is 58 ft. 7 in. long by 14 ft. 7 in. wide ; the side aisle of chancel 41 ft. long by 20 ft. wide ; making the total length of the remaining aisle 107 ft. 3 in. over all. The chapter house measures 44 ft. 10 in. by 11 ft. 8 in. inside. The font is placed below the east window, being evidently removed from the west end when that part was converted into a private

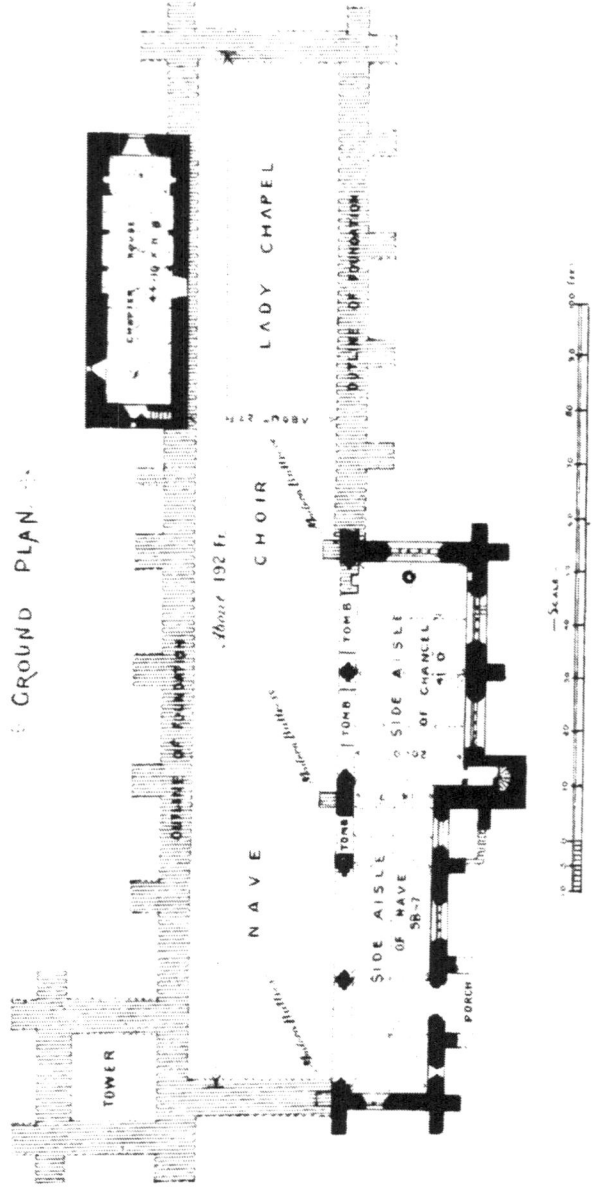

- ⁜ FORTROSE CATHEDRAL ⁜ ·
ROSS-SHIRE

GROUND PLAN.

OUTLINE OF FOUNDATION

About 192 Ft.

NAVE

CHOIR

TOWER

Side Aisle
of Nave
98·7

PORCH

TOMB

TOMB

TOMB

Side Aisle
of Chancel
41 0

CHAPTER HOUSE
44·10 × 21·8

LADY CHAPEL

OUTLINE OF FOUNDATION

ABOUT 213 Ft.

SCALE

burial place ; the position the font now occupies was evidently the site of the altar, the ambry and tercina on either side being conclusive proof for this supposition. The east gable, as well as many other parts of the structure, show irregularity in the levels and deviation from a common centre line of the windows. In the eastern gable this is seen more prominently, owing to the placing of three windows above each other, each window being in the centre of its respective part of the wall. Other peculiarities in the building can be similarly accounted for. We are much indebted to Mr Alex. Mackintosh, architect, Edinburgh, for the above measurements.

There are some curious and interesting inscriptions to be seen about the cathedral ; the following is on a stone built into the wall in a recess beneath the Rood Tower :—

NASCENTES MORIMUR MORIENTES NASCIMUR.

> *As man as soon as born begins to Dye,*
> *So Death Begins Man's Life of Immortality ;*
> *Death, Nature, Time, adieu! all hail, Eternity!*
> *Man's endless state must Be or Happiness or woe.*
> *Tremendous their cause who Strive to show*
> *Annihilation as a safer Creed,*
> *And Mankind a Mulum Necus Breed ;*
> *Were not a Hereafter Man's predestinated lot,*
> *Man's Destiny would Be to Revel and to rot,*
> *Nature's Shame and foulest Blote.*

The following is a translation of a Latin inscription to one Bailie Forbes :—

"In the hope of a blessed resurrection in the Lord, here are placed the ashes of Thomas Forbes, formerly bailie of Fortrose, who died on the 25th and was buried on the 29th of May 1699. He left, as a token of his gratitude to God and his good-will to men, £1200 Scots, to maintain the preaching of the Gospel in this city. His widow, Helen Stuart, hoping hereafter to be buried in this place, has erected this monument to her husband."

On the 21st October 1680 (?) we find letters of mortification by "James, by the mercie of God," Bishop of Ross, Master John Dallas, Dean of Ross, the Chanter, Chancellor, Arch-Deacon, Parsons, &c., of Ross, granting, to the School of Chanonrie and Schoolmasters serving thereat, certain lands, &c., lying in and about Fortrose, being a confirmation of an old mortification granted by the late Bishops of Ross, Dean, and other members of the chapter of the Cathedral Kirk thereof now (" twa hund-reth yeirs past and mair "), of "All and Haill the Feu and Blench duties dew and payable to all and each of them furth of all and sundrie their manses, crofts, houses, and tenements lyand within the said Channonrie of Ross, and about the said Cathedral Kirk of the same, together with the said Church yaird, and grass maill or duty got therefrom . . . . and that as a help for maintaining the Schoolmaster of Channonry."

The cathedral green is now partly used as recreation ground and partly as a grave yard, for though the

> . . . . . " foe
> Hath laid the Lady Chapel low,
> Yet still, beneath the hallow'd soil,
> The peasant rests him from his toil ;
> And dying, bids his bones be laid
> Where erst his simple fathers pray'd."

ROSEMARKIE.

Ancient History.

"The Burgh of Rosemarkie,"* styled by Bishop Leslie about 1578 "a veri ancient town," (the distance between the cathedral at Fortrose and the church of Rosemarkie is just one mile), is said to have been erected a royal burgh, or burgh of regality (?) by Alexander, "King of Scots." In 1255 a charter by Laurence the soldier [miles], witnessed by several of the clergy of Ross, was given at Rosmarc, apparently the burgh.

In 1455 "the toun of Forterose callit the Channonrie of Rose" was annexed by King James II. to the Burgh of Rois-markie . . . . . In 1553 the Queen, on the narrative that the town of Rosemerkie had been of old created a burgh of

---

* Rosmarkyn is supposed to be of Gaelic etymology, composed of *Ros*, signifying a promontory, or headland ; and *maraichin*, seamen : and called by some nomenclatures, "The Point of Ross," being the first land sighted coming up the Moray Firth. (?)

ROSEMARKIE BAY.

regality by her predecessors, and had been annexed to the burgh of the Channonry of Ros, and desirous that the inhabitants should provide for the lodging of strangers resorting thither, granted in favour of David, bishop of Ros, that the bailies, burgesses, and inhabitants of Rosmerkie should have within the Burgh a market cross, a weekly market on Saturday for all kinds of merchandise and wares, and yearly fairs upon Saint Peter's day (1st August) and All-hallow-day (1st November), and on the octaves of both, with power to the bailies to levy all the customs and make payment of them to the bishop.

In 1554 the same queen created the town of Rosmarky a burgh of barony in favour of the bailies, council, and community, the grantees paying yearly to the bishop of Ross the usual burgh fermes, and a wild goose *(anser silvestris),* or the common price of same on the entry of every burgess. . . . . . In 1590 King James VI. created Fortrose, " of old called the Channonrie of Rosse," a royal burgh, with weekly markets on Saturday and Monday, and two yearly fairs, one on Saint Boniface day, and the other on the day called Pardon day (Easter). In 1592 he confirmed the union of Fortrose and Rosemarkie by King James II. . . . . In 1612 King James VI. confirmed the erection of the burgh of Roismarkie, and all the privileges granted to it by his predecessors, Alexander King of Scots and James II. King of Scots, and also the union of the towns by the latter king, uniting them anew, and granting all the privileges of the burgh of Roismarkie (including the fairs on Saint Peter's and All Saint's Days) to the united burgh, which was to be governed by the provost, bailies, and council of the former.

The following is the Latin copy, with translation of precept of Rosemarkie charter, dated 1553 :—

COPY PRECEPT OF THE CHARTER OF THE BURGH OF ROSEMARKIE.

Dated 19th January 1553. Copied from Register of Private Seal, book 26, fol. 39.

" Preceptum Carte ballinonum consulum et Comunitatis burgi de Rossmky facien et creano villam de Rossmarky in liberum burgum in baronia imperpetuum, Reddendo annuatim dicti ballini consules et communitas Reuerendo pri Davidi Rosseu, epso et suis successoribus firmas

decayed, and the houses and buildings thereof become altogether ruinous and demolished, as also dispeopled, there being none but some few resid-ences therein, and most of them all puire fishermen, and that there has been no trade or merchandising within the said burgh this many yeires agoe, nor any courtes keept within the samin for admonition of justice and for punishing of delinquents, trespassers, and malefactors, nor yet any sure place, firmance, or tolbuith therein, wherein they may be secured and incarcerated till Justice have place and be duly execute against them according to the degree of their guilt, conforme to the laudable laws of this realme, in that caise providet, and that the said burgh of Fortrose, formerlie erected in a burgh Royall, and afterwarde incorporate with the said old burgh of Rosemarkie, as said is, Is within ane rig length to the same old and ruinous burgh, and of a most pleasant stance and situation, and of old the cathedral seat of the diocese of Rosse, and a toun, consist-ing of many guid and considerable buildings and houses, and able to afford all kyndes of accomodation, both to the inhabitants thereof and utheres of his Majesty's Leidges resorting thereto. As also given to virtue and dailie increasing, and flourishing more and more in all manner of trade, policie, and industrie. The most part of the inhabitants there-of being merchandis and adventureris."

As a further recommendation of Fortrose, the Act refers to "a most sure and strong firmance, ward hous, and tolbooth for keeping prisoners and malefactors upon all occasiouns."

Rosemarkie has vastly improved since that period, no doubt stimulated and assisted by its union with the flourishing sister burgh.

We append copy of letter to the Bishop of Ross, dated 1st March 1552, intimating the annexation of Rosemarkie, copied from the Register of Private Seals, book 25, folio 55 :—

Ane Lre maid to David Bischope of Ross, makan mentioun yat ye toun of Rossmkie wes infeft and creat of auld in fre burgh of regalite be or Souerane Ladyis predicessouris of befoir @nexit to burgh of ye Chanory of Ross, and sua reput and hauldin in time bygane, and or. souane Lady willi—g to gife occasioun to ye inhabitaries yairof to mak policy wtin ye said to un for lugang of strangis resortant yairto, hes grantit and gevin lettres to ye baillies, burgerss, and inhabitre of ye said burt. To have and hald wtin ye same ane m'cat croce and m'cat day oulkly on Setterday, for selling of all kynd of m'chande and waris within ye sam, and also to have fre faris zerlie vpoun Sact Peteris day and be ye octavis yof, and vpoun alhallow day, and be ye octavis of ye sami, wt power to ye bailies of ye said burgh To intromet and tak vp all custumes of ye saidis faris and inbring and mak payme't yof to David, bischop of Ross, and his succes-sore wt. comand to all and sundrie or sonerane Ladyis vyiris qubatsumevir yat nane of yame tak vpoun hand to mak ony interruptioun, stope truble,

burgales solitas et consuetas ac ad introitum unius cuinsq burgenss unus silvestrem anserem ant eiusdem precium comune datum apud Edr. decimo nono menss Jannary Anno Domini millisimo quingentisso quinquagesimo tertio et regni uri duo decimo.   " PER SIGNETUM."

[TRANSLATION.]

" Precept of the Charter of the Bailies, Council, and Community of the Burgh of Rosemarkie, making and creating the villa of Rosemarkie into a free Burgh of Barony forever. Rendering annually the said Bailies, Council, and Community to the Reverend Father, David Bishop of Ross, and his successors, the Burgh ferms, and use and wont ; and at the admission of every Burgess, one wild goose or the common price of the same. Given at Edinburgh the nineteenth day of the month of January, in the year of our Lord one thousand five hundred and fifty three, and twelfth of our reign.

" PER SIGNETUM."

Change of Name.

We have the following quaint description of how the name of the burgh was changed to Fortrose :—

" In 1661, King Charles II., considering the ruinous state of the burgh of Rosemarkie, then almost depopulated, and the flourishing condition of the burgh of Forterose, which 'is within a rig length to the same old and ruinous burgh, and of a most pleasant stance and cituation, and of old the cathedrall seate of the dyocie of Rosse,' and that the latter still retained its privileges as a royal burgh, was, with the consent of the inhabitants [who from the above facts must have been very few] of Rosemarkie who were to be burgesses of Forterose, confirmed all previous charters and infeftments ; ordained that the united towns should henceforth be called the burgh of Forterose." We find that the markets were again changed, and the half-yearly fairs were each to continue for three days.

In an Act of Parliament, 21st May 1661, confirming the union of Rosemarkie with Fortrose, we have a deplorable narrative of the dilapidated condition of Rosemarkie—

. . . . . " Our soveraigne Lord and his saidis three estates of Parliament, considering that the said burgh of Rosemarkie is now totallie

SEAL OF ROSEMARKIE.

decayed, and the houses and buildings thereof become altogether ruinous and demolished, as also dispeopled, there being none but some few residences therein, and most of them all puire fishermen, and that there has been no trade or merchandising within the said burgh this many yeires agoe, nor any courtes keept within the samin for admonition of justice and for punishing of delinquents, trespassers, and malefactors, nor yet any sure place, firmance, or tolbuith therein, wherein they may be secured and incarcerated till Justice have place and be duly execute against them according to the degree of their guilt, conforme to the laudable laws of this realme, in that caise providet, and that the said burgh of Fortrose, formerlie erected in a burgh Royall, and afterwarde incorporate with the said old burgh of Rosemarkie, as said is, Is within ane rig length to the same old and ruinous burgh, and of a most pleasant stance and situation, and of old the cathedral seat of the diocese of Rosse, and a toun, consisting of many guid and considerable buildings and houses, and able to afford all kyndes of accomodation, both to the inhabitants thereof and utheres of his Majesty's Leidges resorting thereto. As also given to virtue and dailie increasing, and flourishing more and more in all manner of trade, policie, and industrie. The most part of the inhabitants thereof being merchandis and adventureris."

As a further recommendation of Fortrose, the Act refers to "a most sure and strong firmance, ward hous, and tolbooth for keeping prisoners and malefactors upon all occasiouns."

Rosemarkie has vastly improved since that period, no doubt stimulated and assisted by its union with the flourishing sister burgh.

We append copy of letter to the Bishop of Ross, dated 1st March 1552, intimating the annexation of Rosemarkie, copied from the Register of Private Seals, book 25, folio 55 :—

Ane Lre maid to David Bischope of Ross, makan mentioun yat ye toun of Rossmkie wes infeft and creat of auld in fre burgh of regalite be or Souerane Ladyis predicessouris of befoir @nexit to burgh of ye Chanory of Ross, and sua reput and hauldin in time bygane, and or. souane Lady willi—g to gife occasioun to ye inhabitaries yairof to mak policy wtin ye said to un for lugang of strangis resortant yairto, hes grantit and gevin lettres to ye baillies, burgerss, and inhabitre of ye said burt. To have and hald wtin ye same ane m'cat croce and m'cat day oulkly on Setterday, for selling of all kynd of m'chande and waris within ye sam, and also to have fre faris zerlie vpoun Sact Peteris day and be ye octavis yof, and vpoun alhallow day, and be ye octavis of ye sami, wt power to ye bailies of ye said burgh To intromet and tak vp all custumes of ye saidis faris and inbring and mak payme't yof to David, bischop of Ross, and his successore wt. comand to all and sundrie or sonerane Ladyis vyiris quhatsumevir yat nane of yame tak vpoun hand to mak ony interruptioun, stope truble, or Impedimet to ye saidis baillies, burgesss, and comunite in haldin of ye said m'cat wekly vpoun Setterday within ye said burt, and of ye saidis fre faris zerlie at ye feste above wrtti, and be ye octavis yof &ca at Lynlytgw ye first day of M'che, The zeir foirsaid (viz. jm. vc ly o).—Per Signaturam."

The older houses of the inhabitants are chiefly the old residences of the canons. The old seal of the chapter of Ross is now used as the seal of the burgh of Fortrose. The cross of Rosemarkie still stands at the west end of the town, and the seal of the burgh, still in existence, bears the legend SIGILLUM COMMVNE BVRGI DE ROSMARKYN.

ROSEMARKIE CHURCH.

The origin of the Church of Rosemarkie is ascribed to Saint Boniface, surnamed Queretinus, an Italian, who in the seventh or eighth century is said to have come into Scotland for the purpose of inducing the Church there to conform to the practice of the Church of Rome, and, after founding churches in many parts of the country, settled at Rosemarkie, and to have built there a church, in which he was afterwards buried. The church does not appear in any known record, from the period of its foundation till the year 1510 (in which it is mentioned in the "Aberdeen Breviary" as the burying-place of Saint Maloc,) except in "Wynton's Chronicle," where he says :—

> " Sevyn hundyr wynter and sextene,
>   Quhen lychtare wes the Virgyne clene,
>   Pape of Rome than Gregore
>   The secund, quham off yhe herd before,
>   And Anastas than Empryowre,
>   The fyrst yhere off hys honowre
>   Nectan Derly wes than regnand
>   Ower the Pechytis in Scotland.
>   In Ros he fowndyd Rosomarkyne,
>   That dowyd wes whtht Kyngvs syne
>   And made wes a place Cathedrale,
>   Be-north Murrave severalle ;
>   Quhare chanownys are seculare
>   Wndyr Sayant Bonyface lyvand thare."
>   ANDREW WYNTON'S CRONYKIL,

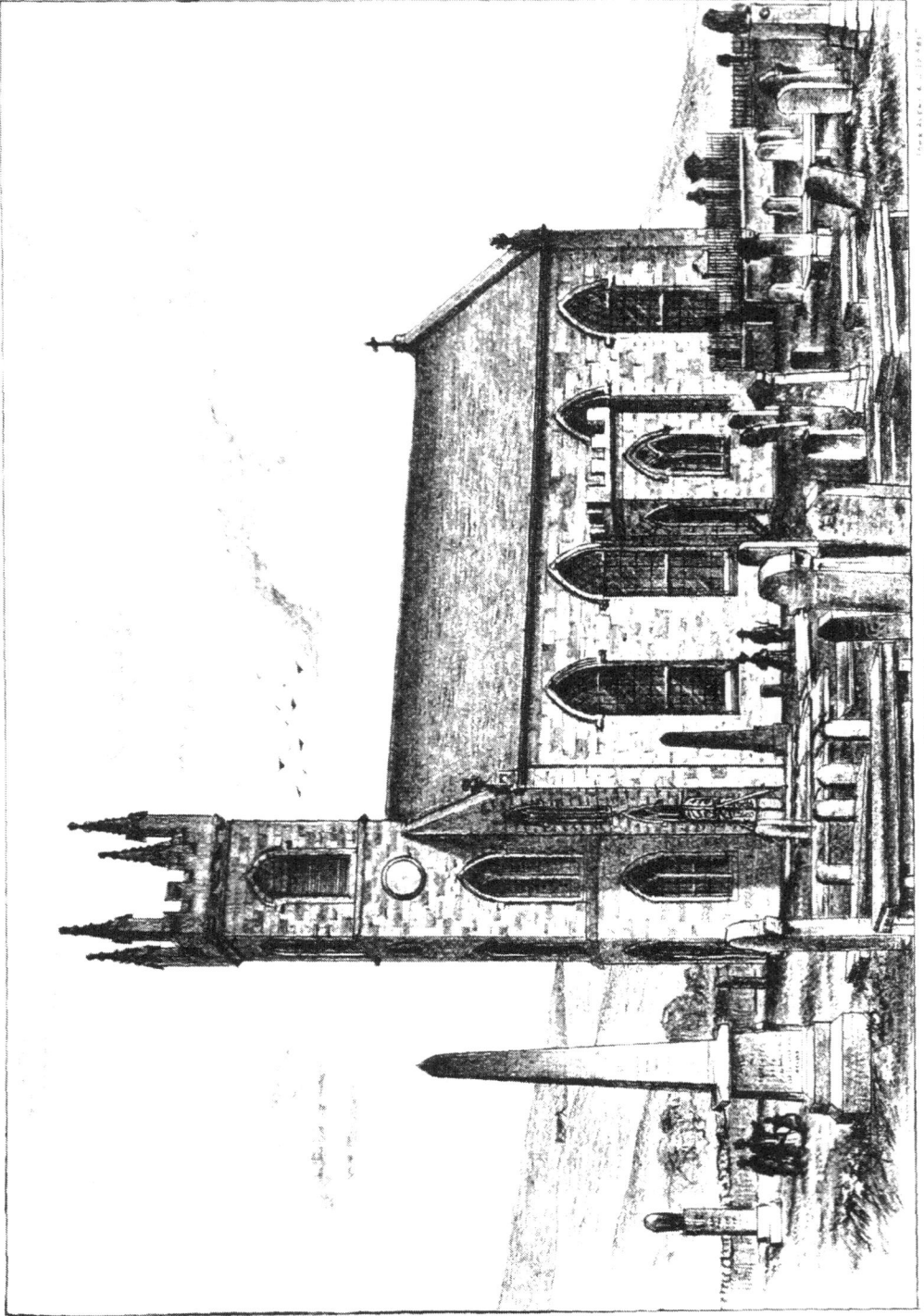

ROSEMARKIE CHURCH.

Maluog or Lagadius, an abbot and bishop of Lismore, who died in 577, is said to have founded a Columban monastery in Rosemarkie. It has been contended that Saint Malrube, Saint Columba's disciple, was buried here, and the Runic cross and ring hereafter described marked his grave. I think, however, it is almost certain that Malrube was buried in Lochmaree island churchyard, and it is very doubtful if he ever ventured to the Black Isle, although he is alleged to have been murdered by the Danes in 722 at Ferrintosh.

In the " Origines Parochiales Scoticæ " we find the following further description :—

"The church of Rosemarky, dedicated to Saint Boniface, stood in the town of Rosemarky, on a bank of sand near the sea shore. In repairing it in 1735, there were found in a vault, under an ancient steeple, some stone coffins of rude workmanship. A new church was built in 1821 on the same site. A well at Rosemarky is still known as Saint Boniface' well."

This well is at Fortrose, not at Rosemarkie, and partly supplies Fortrose with water. For a detailed description of the antiquities and finds made in Rosemarkie and Fortrose, see under heading of antiquities.

Excursions.

The following is a list of excursions that may be made from Fortrose or Rosemarkie as centres, and a general idea can be had by reading the epitomised list, and when any particular excursion is fixed upon, the descriptive chapter can then be perused :—

	Miles.
No 1.—Along Shore Road to Avoch	$1\frac{1}{2}$
Ormond Hill and Castle , ...	$1\frac{1}{4}$
Drive around Ormond Hill, view Munlochy Bay and Valley, Inverness, and Great Glen	$\frac{3}{4}$
Bennet's Monument, Craigack Well, and Sandstone Quarry, out of which Fort-George was built . .	1
Back by Bennetsfield, Rosehaugh House, Valley, Arkendeith Tower, &c.	5
Total miles to walk	$9\frac{1}{2}$

Miles.

No 2.—Pass Mount Pleasant 1¼
    Bog of Shannon, rare plant "Alpina Pinguicula" .   .   1
    Along Road to E. Auchterflow—Bishop's Well, Petri-
    fying stream, and old burying ground   .   .   2¼
    Return 4¾
           Total miles to walk. 9½

No. 3.—Down Chanonry Point, remains of Cross where last witch
    in Scotland was burnt   1
    Across Ferry to Fort-George—the Garrison .   .   .
    Fort-George to Ardersier Village—Beach and clay hills,
    Caledonian Encampment 1½
    Ardersier to Station 1½
    Return journey   4
           Total miles to walk   8

No. 4.—Along Cromarty Road, Rosemarkie—Church and Runic
    Cross—Courthill.   1
    Rosemarkie—Valley and Cliffs of Boulder Clay .   .   ½
    St Helena, the Fairy Glen, and then follow Invergordon
    Road to Blackstan—Site of Battle between Scots and
    Danes—Gray Cairn   4
    Return journey 5½
           Total miles to walk.   11

No. 5.—Along Shore to Eathie Burn, return by Cromarty Road,
    Remarkable Fossil Beds, old Coal Pits, Interesting
    Cliffs and Caves along shore, &c..   .   .   .   7
    Eathie rock section, old Fort, return journey top of cliffs
    or road   6
           Total miles to walk   13

Having epitomised the various walking tours, each of which
may easily be accomplished in a day, we proceed to detail con-
secutively the objects of interest and antiquity in each excursion.

ROUTE No. I.

*Craig Wood, Avoch, Ormond Hill and Castle, Bennet's Monument,
Craigack Well, Rosehaugh House, Avoch House, Arkendeith
Tower.*

Proceeding along the shore road towards Avoch, we pass

AVOCH CHURCH, AND ROSEHAUGH VALLEY

the Craigwood and rocks formerly noticed, and reach the fishing village of Avoch, a mile and a half from Fortrose.

Abh-ach or Avoch, signifying the "Stream of the Field," is situated at the entrance to the Rosehaugh valley, a charming glade, taking its name from the abundance of the little white Burnet rose *(Rosa Spinosissima)*, once so exuberantly numerous and profusely scattered all over the valley, hence the name *Valis Rosarum,* or Rosehaugh.

About two miles up the valley is situated the stately and picturesque residence of Mr Fletcher, who, through enterprise and liberal expenditure on judicial improvements, has enhanced the value of his estates by a third since they came into his hands.

Mr Fletcher has kindly thrown open his private grounds for inspection, and the tourist would be well repaid by a visit.

To Mr Fletcher of Rosehaugh the village of Avoch owes all the improvements made during the last ten years, and these have been considerable ; in fact we may say within that period it has been entirely renovated, good substantial and comfortable houses erected, an abundant supply of water introduced, and sanitary arrangements immensely improved.*　The village, however, can never become fully developed as a fishing station until a proper harbour is erected, and this the proprietor is willing to do should the Fishery Board co-operate with him. The majority of the fishermen are industrious, and their fleet has been greatly augmented of late, but being hampered with this grievance their ambition is too often crushed and success relatively marred.

---

* The following is an extract from a report on Avoch by Mr Peterkin, inspector of the Board of Supervision :—"The effect of the present arrangement is that the village of Avoch, from having been one of the most slovenly and filthy, has now been transformed into one of the most tidy and clean fishing villages to be found anywhere. . . . . There can, I think, be but one opinion held by those who formerly knew Avoch intimately, and who now see it—that a marked sanitary improvement has been effected. Instead of dirty lanes, tenanted by pigs—instead of dung-pits, dung-heaps, and scattered domestic refuse, ashes, &c., being found throughout the village, every lane and street will be found to be in good condition, and clean and free from all these things. Large boxes are conveniently placed throughout the village for the reception of domestic refuse, the contents of which are regularly taken away by the scavenger, and carried by him to the depot, to be removed frequently o a distance.

A writer eighty years ago says—"At Avoch we find them manufacturing coarse linen from lint raised at home; herring and salmon nets, and fishing tackle, partly from hemp raised there also; and upon the bold and adventurous fishermen of Avoch do the Invernessians chiefly depend for their supply of fine fish." This industry, as in many other parts of the North, is now dormant; and although Avoch possesses woollen mills, we regret that the business is so meagrely developed that its presence little benefits the place.* The firth opposite is well adapted for oyster beds, and was until recently the first bank in the North, but by wholesale dredging, or trawling, and mismanagement they killed the goose which laid the golden eggs. Were an enterprising company to undertake the establishment of an oyster farm, and thoroughly develop the banks, there is little doubt the speculation would be remunerative to both the shareholders and fishermen. If the Avoch fishermen undertook the cultivation of the beds, the Government or Fishery Board would most probably give them exclusive right of fishing them. Avoch has three elegant places of worship, a library and reading room, a public school, and the Mackenzie Institute, an endowed school, under the management of the Episcopal diocese.

Ormond Castle.

About a mile to the west of Avoch are the remains of what was once one of the strongest castle residences in the North, called Ormond Castle. It is supposed to have been one of the many royal castles erected in the twelfth century for holding in awe the disloyal of the inhabitants. We find it chronicled that King William the Lion erected in 1179 two castles in the Lordship of Ardmanach (the Black Isle). One of these was Redcastle, and Ormond Castle is supposed to be the other.

Before noting our own description and notes on the ruins, we will make a few quotations from early writers. In the "Ori-

---

* It is a lamentable fact that the north of Scotland has retrograded during the last half century in its woollen and other manufactories of native produce. Possibly their re-establishment might do much to alleviate agricultural sufferings and depression of local trade.

gines Parochiales Scoticæ," we have the following interesting history :—

"From the Castle of Avoch, known as the Castle of Ormond, Ormondy, or Ormond Hill, and Douglas Castle, Hugh of Douglas, between 1440 and 1448, drew the style of Earl of Ormond, and James Stewart, the second son of King James III., between 1460 and 1481, drew the style of Marquis of Ormond. In 1481, as we have seen, King James III. granted the lands of Avauch, with the moothill of Ormond, to the Marquis of Ormond, who, about 1503, resigned the lands, but retained the moothill in order to preserve his title. A writer of the seventeenth century mentions Ormond Hill south west from the church [of Avoch], with the remains of a castle, and elsewhere describes it as Castletown, with the ruynes of a castle called the Castle of Ormond, which hath gevin styles to sundrie earls, and last to the princes of Scotland. The foundations of the castle remain on the top of a hill near Castletown point on the Bay of Munlochy, about 200 feet above the level of the sea. They occupy a space 350 feet by 160 feet, and the castle seems to have been built of coarse red sandstone and lime, with a ditch on one side. The Hill of Castletown is now known as Ormond Hill or Ladyhill (the latter name having arisen evidently from the dedication of its chapel)."

The following is from an extract quoted in *Anderson's Guide to the Highlands* :—

"On a rocky mound, now called 'Ormond,' or 'Lady Hill,' at the west end of these green links [referring to the green sward that formerly extended along the sea-shore between Avoch and Fortrose] stood the ancient castle of Avoch, to which, as related by Wyntoun, the Regent Sir Andrew de Moravia, 'a lord of great bounty, of sober and chaste life, wise and upright in council, liberal and generous, devout and charitable, stout, hardy, and of great courage,' retired from the fatigues of war, and ended his days about the year 1338, and was buried in the 'Cathedral kirk of Rosmarkin.'* Passing afterwards into the possession of the earls of Ross, this castle was, on their forfeiture in 1746, annexed to the Crown, when James III. created his second son Duke of Ross, Marquis of Ormond, and Earl of Edirdal, otherwise called Ardmanache, and hence this district, which still bears these names, thus became one of the regular appanages of the Royal family of Scotland."†

In the annexation in 1455, Redcastle, with the lordship of Ross pertaining thereto, is particularly mentioned ; and mention is also made of the house of Innerness and Urquhard

* Fraser Tytler's History of Scotland, vol. II., page 65.
† The earldom of Ross and lordship of Ardmanach are appointed to be the patrimony of the King's second son. Jas. 6, par. II., cap. 30.

on Loch Ness, and of "Annache (Avoch) Edderdail, callyt Ard-manache." And this annexation in the time of James II. was repeated and confirmed by the whole Parliament on the 1st July 1476, in favour of James III., who afterwards, on 29th January 1487, created his son Duke of Ross, *Marquis of Ormond*, and Earl of Edirdal, otherwise called Ardmanache ; from which period the lordship of Ardmanache, or the Black Isle, was gene-rally considered as part of the patrimony of the king's second son (see Act Parliament, Scot., Thomson's folio edition, pp. 42, 113, and 181).

Robert Young, Esq., Elgin, writing on the subject, says—

"The castle of Avoch stood on the Ormond Hill, and commanded an extensive view both of land and sea, particularly of the upper reach of the Moray Firth, and of the counties of Moray, Nairn, and Inverness. It now forms a part of the estate of Rosehaugh, belonging to James Fletcher, Esq., who, among his other extensive and varied improvements, as well agricultural as ornamental, has lately planted the Hill, and formed a carriage drive around it. Some of the foundations still exist, and it would seem that the building had been an extensive one. Sir Andrew Moray was married to Christian Bruce, sister of King Robert Bruce, and, being appointed Regent of Scotland during the early part of the reign of David Bruce, he took a most prominent and successful part in the defence of the kingdom at that very perilous period of its history. Next to Bruce and Wallace (and not even excepting Douglas and Randolph), we owe the independence of Scotland (under providence) more to Sir Andrew Moray than to any other man. The castle of Avoch should therefore be esteemed a hallowed spot by every true-hearted Scotsman. The Ormond Hill is said to have given the title of Ormond to Hugh Douglas, brother of the Earl of Douglas, who was raised to that rank about the middle of the 15th century. He only enjoyed it for a few years, having joined his bro-ther's, the Earl of Douglas, rebellion against his sovereign, he was justly forfeited. It is probable from this circumstance that the Douglases, as successors of the family of De Moravia, had for some time possession of the castle of Avoch and adjoining lands. The rev. author of the new statistical account of the parish of Avoch, in his very interesting sketch, is of opinion that the stones of the castle of Avoch were carried away by Oliver Cromwell to assist in the erection of the Citadel at Inverness, called Cromwell's Fort. This is extremely probable, as stones were taken for that purpose wherever they could be found, without any scruples, as for instance the case of the Abbey of Kinloss, where the extensive build-ings were almost razed to the ground."

In October 1883, I made a survey of the top of Ormond Hill, but notwithstanding the assistance of several men kindly sent

by Mr Fletcher of Rosehaugh for excavating, I found great difficulty in tracing the original outline. The walls are overgrown
with turf, and to get a thoroughly reliable ground plan, the whole
area must be excavated. So far as measured we found distinct traces
of walls, but the towers, which appear as circular, may be square
if cleared out, as did one which, before being excavated, presented
a circular appearance, but, when dug to a depth of 6 ft., revealed
an inside opening of 9 ft. square, with substantial walls 4 ft.
thick, built of sandstone in mortar. In this tower we found small
stones and mortar; also, a very fine sandstone door rybat, with
the "droved" marks of chisel distinctly visible. This stone is in
the possession of Mr Douglas Fletcher, at Rosehaugh House.

The castle was of an oval shape, following the contour of the
hill, yet strangely enough of identical form to many prehistoric
places of defence and vitrified forts. A wall about 4 feet thick
runs along the top of the slope all round, while on the east side
a fosse traverses the nose of the hill, running for 30 or 40
yards around north and south sides. This ditch is now 6 feet
wide at bottom, with a mound on the east side rising 5 feet
above the bottom, and measuring 9 feet wide on top. All the
slopes about the ditch are about $1\frac{1}{2}$ to 1. The total length from
east to west is about 160 ft., and breadth between walls about
80 ft.

An excavation scooped out in the solid conglomerate rock is
supposed to be the "well," which supplied the castle with water,
but this seems improbable. Being filled with stones, I never
could ascertain its actual depth. Tradition has it that into
this well the treasures of the castle were thrown, and the
building set on fire by the hands of its owner, on seeing the
approach of Cromwell's army against it. That the materials of
the building were removed from the site there is abundance of
evidence, for from the summit to the shore the track where the
stones were rolled down can be easily traced, and a large block
of the building still lying on the shore, and weighing fully 4 tons,
testifies to the truth of this supposition. The block referred to
is composed of small pebbles embedded in lime, and it takes a
hard struggle to extract even one small stone from the mass.
This is surely a tribute to the excellent builder of the remote
age, when might was right, and strength the only logic used.

The sub-marine botany of the firth below the castle is very rich, and rare specimens may easily be fished up.

Leaving the old castle behind, we descend to the drive again, and follow round the hill. The view obtained here is truly grand. Away up at the head of the firth stands the town of Inverness, with its finely wooded slopes and villa studded suburbs, while backwards extends the Glen-more-albyn, with its rugged slopes on either side gradually becoming less distinct, until they seem to recede into and mingle with the pale blue atmosphere. To refer to Inverness is a digression; yet of so lovely a town, and our northern Highland capital, we cannot refrain from quoting Professor Blackie, the "Highland Patriot's" song in its praise :—

> " Some sing of Rome and some of Florence ; I
> Will sound thy Highland praise, fair Inverness :
> And, till some worthier bard thy thanks may buy,
> Hope for the greater, but spurn not the less.
> All things that make a city fair are thine,
> The rightful queen and sovereign of this land
> Of Bens and Glens and valiant men, who shine
> Brightest in Britain's glory-roll, and stand
> Best bulwarks of her bounds—wide-circling sweep
> Of rich green slope and brown-enpurpled brae,
> And flowery mead, and far in-winding bay,
> Temple and tower are thine, and castled keep,
> And ample stream that round fair gardened isles
> Rolls its majestic current, wreathed in smiles."

Looking up the Munlochy Bay the scene is equally fine and picturesque. The still waters of the bay reflect with the clearness of a mirror the varied scenery of its shores ; from the rugged outlines of Craigiebow and Druim-na-deur, to the dark waving pines of Ord Hill and the Scope ; while far up the valley to Allangrange we have a picture varied as it is beautiful ; there we have winding hedgerows, pleasantly relieved with the different hues of cultivated fields, with here and there clumps of variegated foliage blending harmoniously with the grim, dark Gallow-hill filling up the back ground, which at intervals reveals faint glimpses of the blue outlines of the conical mountain peaks of Strathconan and Strathglass.

After semi-encircling the hill, we strike to the left, along the terrace above the bay, until we arrive at Bennet's Monument, an obelisk of sandstone, erected on a mound a little above the road. It forms so conspicuous a feature in the landscape that further detail of route thereto seems unnecessary, yet some remarks on traditions associated with it may prove interesting. In the good old days, "lang syne," when the Black Isle was the undisturbed haunt of giants, witches, and fairies, a man named Bennet was, by the powers that were on the mainland, banished to the Black Isle; whatever misdemeanour poor Bennet committed, suffice it to say the transportation to the Black Isle was considered the most severe punishment that could be inflicted. Bennet, after looking around, squatted down at this place, but what little stock was sent with him for his sustenance soon found a speedy removal, for immediately opposite is the cave of Craigiehow, in which dwelt a giant, who soon found out his new neighbour, and particularly his bullocks, for whenever he felt disposed he waded across the bay, slung an ox across his shoulders, and returned to his abode to have a quiet regale on his stolen booty. It was on one of those unlawful visits that Bennet's monument was erected, the giant growing portly over his sumptuous feasts needed something to assist him up the hill, so picking up this stone slab in the bay, he utilised it as a walking stick, and on reaching the top of the terrace stuck it on end, where it remains until this day, a testimony proving that "giants lived in the land in those days."* Bennet, however, was revenged on the giant by inducing him to commit a suicidal experiment, under the pretence of increasing his strength. Immediately below the monument is Craigack well, greatly resorted to upon the first Sunday of summer, when the waters are supposed to have special virtues. Even in the enlightened age of the 19th century this superstitious practice is largely indulged in, for on visiting the well on the day referred to, the bush above the well appears literally covered with rags of every hue (because some token must be presented to the water nymph, otherwise the virtues of

---

* The monument is a sandstone slab 12 inches by 7 inches, and stands 8 feet 8 inches above a circular mound of about 44 yards diameter, and 10 feet high. The stone bears the date 1752, with initials G.M.K. on one side, and I.M. E.M.K., 1755, with the *Caberfeidh* (stag's head) on the other.

the well are withheld), and other suspicious remains lying about indicate that something stronger than water was indulged in. Adjoining the well is a fine sandstone quarry, from which the greater part of the stones used in building Fort-George was taken. Following the drive again, we pass near the farm of Bennetsfield, said to be the richest, and certainly the finest land in the Black Isle, it being supposed it was the first land tilled in the peninsula.

We now pass by Corrachie farm, and are soon on the public road leading from Inverness to Fortrose. A detour can be made to Rosehaugh and grounds along the private road through the valley and joining the public road again at Avoch. The objects of interest to be seen along the valley road are Rosehaugh House and grounds, which Mr Fletcher has kindly thrown open to visitors who may at all times ramble freely through any portion of his splendid pleasure grounds and magnificent gardens. The Mansion House is a fine specimen of classical work, and the situation is one of the finest in the Black Isle. From the views appended, one can judge the scenery to be fine, but a visit to the spot is the true medium to obtain the conception of the beauty of the place. Midway between Avoch and Rosehaugh, the old house of Avoch, destroyed by fire, is noticed on the left, while on the terrace above are the ruins of a tower at Arkendeith called "*Airc-Eoin-dubh*," Black John's Ark, or place of safety. Black John was a Highland reiver, making raids on his surrounding neighbours, and carrying his booty in safety to his strong fortaliced dwelling. Nothing now remains but the lower story, consisting of a strong walled square building, with the arched roof of the dungeon still intact. It seems a contemporaneous erection with Fairburn Tower, in the west of the Black Isle, and perhaps not of such antiquity as is ascribed to it. In the printed retours for the seventeenth century (1611-18) is a special mention of some of the Bruces of Kinloch holding the lands of Muirale House and Arkindeuch.

Below Arkendeith Tower are the ruins of Avoch House, destroyed by fire ; it was built partly after the Scottish baronial style, and from its pleasant situation deserves a visit.

As we pass again through the village of Avoch, we notice the strange dialect, peculiar to the fisher community, while in all this parish and the greater parts of Rosemarkie and Cromarty

ROSEHAUGH HOUSE

(The Seat of James Fletcher, Esq.)

parishes Gaelic is unknown. This is ascribed to these parts being settled by Scandinavians.

EXCURSION No. II.

AUCHTERFLOW DISTRICT.—*Mount Pleasant, Bog of Shannon, Strath of Auchterflow, Bishop's Well, Petrifying Stream, Old Burying-Yard, Sandstone Quarry, Rare Plant.*

This district, although presenting little or no attraction to the antiquary, will be of intense interest to the geologist, and particularly to the botanist, because the rare plant *Pingicula Alpina*, which has not yet been found in any other part of Great Britain, is found in profusion here.

The road to be followed to reach the Strath of Auchterflow leads behind the town, passes Mount Pleasant, Bog of Shannon, and Killen. From this point a magnificent view is obtained. The Strath presents the appearance of a plateau, extending westward for three miles, and the same distance to the east; the geological study of which is interesting. There is every indication that this plain was covered by a fresh water lake, for marl, peat, and decayed vegetable matter are found near Raddery and Bog of Shannon. The School of Killen, where we stand, was erected in 1868 by Mr Fletcher of Rosehaugh, previous to the introduction of School Boards, chiefly for the convenience of his tenantry's children. The school is now under the control of the Board, but, previously, Mr Fletcher contributed liberally to the teachers' salaries, and even yet evinces much interest in its welfare.

The only objects of antiquarian interest are:—A spring called the "Bishop's Well," at Shawpark, in the west end of the Strath, and an ancient burying-ground, said to be also the site of a chapel, near the Killen burn, a little below Bog of Afterflow farm. Between this farm and the burying-ground is a petrifying stream, trickling over the embankment, and some fine specimens of delicately petrified moss, ferns, &c., may easily be found. At this point of the burn of Killen, Hugh Miller discovered an ichthyolitic bed, which yielded to his hammer specimens of the *Diplacanthus, Striatus, Cheiracanthus,* and several species of the *Coccosteus.*

From this region eastward by Boggiewell to Raddery, a dis-

tance of about three miles, the *P. alpina* was traced by the Rev.
G. Gordon of Birnie in 1831, his attention being called to it by
Mr Campbell Smith, land-surveyor, in June of that year.   The
parish of Avoch contains as rich and extensive a flora as any in
Scotland, the clayish quality of the soil and debris of decomposed
granites produce numerous native herbaceous plants.

In the Statistical Account of Scotland, page 388, we find
stated that " In no part of the Highlands are more luxurious
festoons to be seen of *vicia sylvatica,* or larger or more showy
specimens of the *Geranium sylvaticum,* *G. sanguineum,* and
Saxifraga granulata.   The roses which make such a show in
the same neighbourhood, and which caused the celebrated courtier,
Sir George Mackenzie, to style his property 'Rosehaugh' are of
the species *Rosa Canina* and *R. Spinosissima.*"

The return journey may be continued down the Killen Burn
to Rosehaugh, or an equi-distant tour may be made by Raddery,
and back by the Hill of Fortrose.

EXCURSION No. III.

Chanonry Point, Remains of Cross and Traditions, Ferry, Fort-
George Garrison, Campbelltown, Clay Hills, Remains of
Caledonian Encampment, Cawdor Castle.

This excursion is very different from any of the others ; and
though offering but tame scenery, will form a break in the monotony of uniformity.   The Chanonry Point, with its fine "links,"
smooth as a carpet, and admirably adapted for the popular game
of golf, runs out for 1 mile 1100 yards from Fortrose, and will
present an interesting problem to the geologist to account for the
formation of this and Fort-George promontory.   Chanonry Point
is composed chiefly of fine sand, while the opposite one presents
masses of rough shingle ; how the various flows of strong currents have thus deposited the sand on one side and shingle on
the other, should receive the attention of the skilled geologist.*

---

* In the bay opposite the Episcopal Church is a good mussel bed,
which the Council might turn to advantage, and adjoining the scalp is a
peculiar rock, called Craig-an-roan.   It is now always dry at low water,
while many years ago it was seldom visible.   It seems to be a concre-
tionary rock, gradually forming of minute shell and depository animal-
culae.   It certainly deserves investigation.

There is no peculiarity in nature without its tradition, so with those points. The good fairies who inhabited the Black Isle were ordered by the arch-wizard Michael Scott, who was then receiving education at Fortrose, to make a rope of sand. The little "people in green" set a-spinning their problem, making rapid progress so as to complete " ere the muir cocks 'gan to craw," and would have reached the opposite shore, did not one of those still to be met with too appreciative and premature complimentary individuals, who had to cross the ferry, seeing the works rapidly spanning the firth, bless them in anticipation of saving his ferry rate. This untimely meddling abruptly terminated the work, and " they've never done anything since."

Eleven hundred and thirty-three yards north of the Lighthouse, and one hundred and sixty-six yards east of the public road, is the stump of a cross where tradition asserts that the last witch in Scotland was burnt. The Chanonry Point seems to have been the common cremation ground, for it is related that here Epack of Ord Hill, the Munlochy Bay witch, and the Brahan Seer, were burnt, each in a barrel of tar. Why they were executed here is clear: the Church was the prosecutor, and under the eye of its dignitaries poor innocent creatures were cruelly murdered if they presented any peculiar deformity, either mentally or physically.

Fort-George Ferry, which is 1433 yards wide, is manned by safe boats, and although the current during spring tides is rapid, no alarm need be apprehended. The deepest part, according to the Admiralty chart, is 154 feet. On arriving at the garrison, two hours may be spent with interest and profit in examining the ramparts, bombproofs, ditch, and drawbridges.

Garrison.—The garrison was built three years after the rebellion of 1745, at an actual cost of £160,000; it covers 12 acres of ground; has a polygonal line, with six bastions; it is mostly surrounded by water, and is defended on the land side by a ditch, a covert way, a glacis, two lunettes, and a ravelin. From the shingly nature of the foundations, the ramparts have sunk considerably; so much so, that originally no part of the houses were visible from outside, while nearly all the roofs may now be seen. It contains accommodation for 2180 men, although rarely half that number is regularly stationed there. It now forms the headquarters of the Seaforth Highlanders (72nd and 78th regiments), the depot of the Camerons being transferred to the new barracks at Inverness, after the territorial arrangement of the army. About three-quarters of a mile further south, along the shore, is the village of Campbelltown, or Ardersier, where nothing worthy of note is to be seen; but on the ridge rising behind the village are the supposed remains of a Caledonian fort, called Cromal, or Cromwell's mount. In the clay at the bottom, Mr T. D. Wallace, Inverness, discovered Arctic shells; while some miles south westward, up Strathnairn valley, a similar bed of blue clay, containing remains of Arctic shells, was recently discovered by Mr James Fraser, C.E., Inverness, at a height of 500 feet above sea level.

The visitor on going South may return by this route, Fort-George Station being only three miles from the garrison, thus saving the detour by Inverness.

By driving, Cawdor Castle may be visited, made so famous in the tragedy of Macbeth, in which King Duncan was supposed to have been murdered; it is about twelve miles from the village, where vehicles can be hired. The return journey must be practically over the same ground.

EXCURSION No. IV.

Rosemarkie Church, Runic Cross, the Dens and Fairy Glen, Boulder Cliffs, St Helena.

Passing through Rosemarkie, we will not repeat what is referred to under the heading of Rosemarkie and in the appendix on antiquities.

Immediately on leaving the town we observe the huge cliffs of boulder clay which recede from the shore far up the glen. The view looking northwards from here is grand, for Rosemarkie Burn is one of the most charming scenes in the vicinity, with its red weather-worn cliffs of boulder clay, the denudation of many a winters' blast, carving this huge mass into a thousand fantastic forms; here towering for two hundred feet to a pinnacle capped by a solitary boulder; there to perpendicular cliffs perforated, with many a swallow's nestling abode, like the canon-shattered battlements of some ancient castle; anon receding in deep winding gullies, formed by the wintry torrent of many ages; while in the back ground, in summer, the profuse variegated foliage forms a scene well worthy of its name, " The Fairy Glen."

> " Boon nature scatter'd, free and wild,
> Each plant or flower, the mountain's child.
>
> The primrose pale and violet flower
> Found in each cliff a narrow bower.
>
> Foxglove and night-shade, side by side,
> Emblems of punishment and pride,
> Group'd their dark hues with every stain
> The weather-beaten cliffs retain,
> With boughs that quaked at every breath,
> Grey birch and aspen wept beneath,
> So wond'rous wild, the whole might seem
> The scenery of a fairy dream."

Apart from its landscape beauty, the glen offers to the geologist an interesting and puzzling problem; how such a vast accumulation of glacier debris could be deposited in this particular locality, with no indication of the main flow of glaciers in this direction, is difficult to decide. The hard and often regular stratification seems to point to a deposition in apparently still waters. We have often wondered that Hugh Miller never seemed to have given it his minute attention, nor do the geologists of our own day trouble themselves to solve this intricate question. Although Miller never ventures a solution of the boulder beds of the valley, he gives the following graphic description of its scenery :—

" Rosemarkie, with its long narrow valley, and its abrupt red scaurs,* is chiefly interesting to the geologist for its vast beds of boulder clay. I am acquainted with no other locality in the kingdom where this deposit is hollowed into ravines so profound, or presents precipices so imposing and lofty.

It presents the appearance of a hill that had been cut sheer through the middle from top to base, and exhibits in its abrupt front a broad red perpendicular section of at least one hundred feet in height, barred transversely by thin layers of sand, and scored vertically by the slow action of the rains. Originally it must have stretched its vanished limb across the opening, like some huge snow-wreath accumulated athwart a frozen rivulet; but the incessant sweep of the stream that runs through

* SCAUR—Scotice—a precipice of clay. There is no single English word that conveys exactly the same idea.

the valley has long since amputated and carried away ; and so only half the hill now remains. . . . The clay presents here, more than in almost any other locality with which I am acquainted, the character of a stratified deposit. . . A little higher up the valley, on the western side, there occurs in the clay what may be termed a group of excavations, comprising a piece of scenery, ruinously broken and dreary, and that bears a specific character of its own which scarce any other deposit could have exhibited. The excavations are of considerable depth and extent— hollows out of which the materials of pyramids might have been taken. The precipitous sides are fretted by jutting ridges and receding inflections that present in abundance their diversified alternations of light and shadow. . . . Viewed by moonlight, when the pale red of the clay where the beam falls direct is relieved by the intense shadows, these ex- cavations of the valley of Rosemarkie form scenes of strange and ghostly wildness : the projecting, buttress-like angles, the broken walls, the curved inflections, the pointed pinnacles—the turrets, with their masses of projecting coping—the utter lack of vegetation, save where the heath and furze rustle far above—all combine to form assemblages of dreary ruins, amid which, in the solitude of night, one almost expects to see spirits walk."

The only explanation we ever noticed regarding this wonder- ful deposit was that cross currents met somewhere between Fortrose and Fort-George—one moving through the Rosemarkie valley, the other meeting it from the Beauly and the Ness— and the daily ebb and flow of the tide intensifying the conflict of the waters.

It is not within the province of a guide book to propound theories or solve intricate problems, but we may be pardoned for modestly advancing our own convictions on this point. After studying the contours of the district, it seems hardly possible for a current of any great volume to have found its way down the Rosemarkie valley Is it not possible that this deposit was connected with the clay hill on the Ardersier side? Forming a barrier across the firth their union would be of suffi- cient elevation to send the Beauly waters down by Muir of Ord (where there is abundance of evidence to substantiate this sup- position), and discharging the Ness valley drainage westward into the Atlantic ocean. This may account for the well-defined ter- races of the inner firth, while from here down to Cromarty no trace of such is visible. On studying the map it can be readily grasped how strong currents, rushing down the firth (after the barrier was burst), and striking the Campbelltown point or bay, would be detoured, and eddied back in a parabolic sweep, would form the Chanonry promontory. This will immediately explain how the rougher material is deposited on the Ardersier side : the heavier matter first settling, the finer particles, carried by the impetus of the eddy, would gradually settle on the opposite side, or at the theoretic point of where the speed of reversed current would become exhausted.

Having climbed to the height, we reach a point called St Helena, which, from its elevated and romantic position, is greatly resorted to as a pic-nicking ground. The panoramic view of mountain, glen, and sea, observed from here, must be seen to be fully realised. On a clear day the surrounding country can be viewed for miles in all directions.

Following along the road leading to Cromarty for 3½ miles, we notice on north side of road, nearly opposite Glen Urquhart Farm, a huge cairn of stones called " Cairn Glas," the grey cairn. All along the summit of the ridge westward for three miles are numerous cairns and mounds, supposed to mark the site of a battle fought between the Danes and the Scots. In one of the cairns opened in 1811 human remains were found. In the statistical account it is stated that under a cairn called " Cairn-a-Chatb," signifying the cairn of the battle, stone coffins and weapons of copper and other metals were found, confirming the tradition that a huge Danish chief was buried there. " From its nearness to the inviting landing-place of Cromarty it is easy to conceive this district to have been the scene of incessant strife between the northern rovers and the tenacious Gaelic tribes of the country." In 1787 several Elizabethan silver coins were discovered in a small cairn of stones between here and Rosemarkie.

EXCURSON No. V.

Eathie Burn, &c.

The Burn of Eathie, made so famous by the valuable discoveries and lucid illustrations of Hugh Miller, is distant only seven miles walking along the shore, where

"The sleepless billows on the ocean's breast
Break like a bursting heart, and die in foam,"

but to those whose pedestrian powers are undeveloped, a sail on a fine day would be more enjoyable. To the geologist this excursion will be the most interesting of all, and with the "Old Red Sandstone" in his hand he will have no difficulty in noting the many rare formations in this small area, so clearly described by Nature's own geologist—Hugh Miller. Even the most casual amateur in the science will have no difficulty in picking up on the shore many fine specimens of rocks and fossils of the secondary strata, found in patches in the bed of the firth, and along the coast to beyond Helmsdale. Those that abound most on the beach are belimnites, called by the natives thunder-bolts, from their dart-like pointed outline, and ammonites varying in size from two inches to two feet in diameter. The latter fossil is capable of high polish, and may be usefully intro-duced in jewellery adornment.* The best route to reach the burn is following along the shore from Rosemarkie. Low water is the best time to go, as the receded tide exposes washed-up fossils from the patch of oolite referred to.

All along this coast is famed for its excellent salmon fishing grounds. The prevalent rocks between Rosemarkie and Eathie is granite gneiss, and a peculiar granite with the quartz varying from a milk white to a blood-red colour. Mr J. W. Judd, on his map showing the distribution of secondary strata around the shores of the Moray Firth, shows the Lower Silurian extending westward two miles from Eathie Burn, and both the Sutors of Cromarty as belonging to that formation. I have picked up in some parts a peculiar quartzite, which gave a metallic ring when struck, coloured a bluish tint, and had small cubical iron pyrites studded all over. In the same locality I found a solid lump of blue mineral, similar to compressed clay, and pronounced by Dr Aitken, Inverness (a good mineralogist and geologist), as "Abri-achanite." The point where those specimens were picked up was about three miles from Rosemarkie, where the cliffs tower to a great height, presenting a grand and impressive view. All along the face of the cliffs is perforated by caves, and good speci-mens of stalactites may be found in several.

* A gentleman belonging to Fortrose recently found an ammonite below Craighead 38 inches in circumference, and weighing 31¼ lbs.

The coast all along this side of the firth is bold, rocky, and romantic, and abounds in frightful precipices, where

> "Low-browed rocks hang nodding o'er the deep."

From the north Sutor to Tarbetness, in the cliffs may be found granite gneiss, coarse sandstone conglomerate, lias shales, lime-stones, calciferous sandstone, bituminous and calcareous shales; while from the Dornoch Firth to Helmsdale, the upper, middle, and lower oolite fringe the coast; and on the opposite side of the firth, at Burghead, the keuper beds of the trias formation occur.

Garnets are plentifully found in the rocks along the shore. The Hill of Cromarty is noted for the tradition of Wallace encamping here, and destroying the castle. The hill is composed of granite gneiss and splintery hornblende; it has many fine caves, and five vast natural archways in the rock.

Near the Eathie fishing station the rocks recede from the shore line, and a flat plain intervenes. Here some unfortunate speculator sunk two shafts in the hope of discovering coal, but his geological knowledge was less scanty than his revenue when he began, although matters were diametrically reversed ere he finished; for down through the bituminous shale of the lias he worked in ardent expectations, until he met the old red sandstone, which sealed the doom alike of his fortune and fame. The poor unfortunate speculator lost all his worldly means, yet the sympathetic natives raised a public subscription to help him to return to his native land, a sadder and a wiser man. This proved the death knell of coal-searching in the Black Isle. Around the mouths of the pits may be picked up the liassic shaley matter, with thousands of minute fossils imprinted in white on its dark glossy surface.

The Burn of Eathie is but a few hundred yards further a-head, and we can do no better than refer the reader to Hugh Miller's "Old Red Sandstone," where its geology and wild grand scenery are pourtrayed in language so sympathetic and harmonious with what we view, that we seem to feel the pathos that stirred the breast of him who sung its praises many long years ago. We feel it would be doing an injustice to his revered memory were we even to attempt any supplementary description of the Burn of Eathie. The fossils found here are the *coccosteus, pteriehthys*, and *glyptolepis*.

On the summit of the west cliff are the ruins of some ancient structure, which seems to have escaped the scrutinising observation of Miller, although he graphically describes the site or ruins of " Bennet's Chapel," a little further east, where honest Donald Calder, on his homeward journey (probably a market night, and rather late before he left the fair), had such an encounter with the fairies (see " Old Red Sandstone").

The homeward journey may be made along the top of the cliffs (which is rather dangerous walking), or by following the course of the burn until the Cromarty road is reached at Glen Urquhart Farm, when the " Grey Cairn " will be passed, and the Blackstand formerly referred to.

The " Grey Cairn " is at the eastern extremity of what is supposed to be the site of the " Blackstand " battle between the Danes and the inhabitants. The cairn is an aggregation of loose stones, and undoubtedly some important event in a remote age caused this huge though rude monument to be erected. Many years since another cairn existed, but was removed by farmers for building purposes, when human bones were discovered, one skull being so gigantic, according to the statistical account, " as to contain 'two lippies of bear,'" [malted barley.] Along to the top of the moor are hummocky mounds, resembling, and said to be, the graves of those who fell at this apparently sanguinary battle.

Nothing further of interest need here be noted, as the remaining portion of the way is described in route IV.

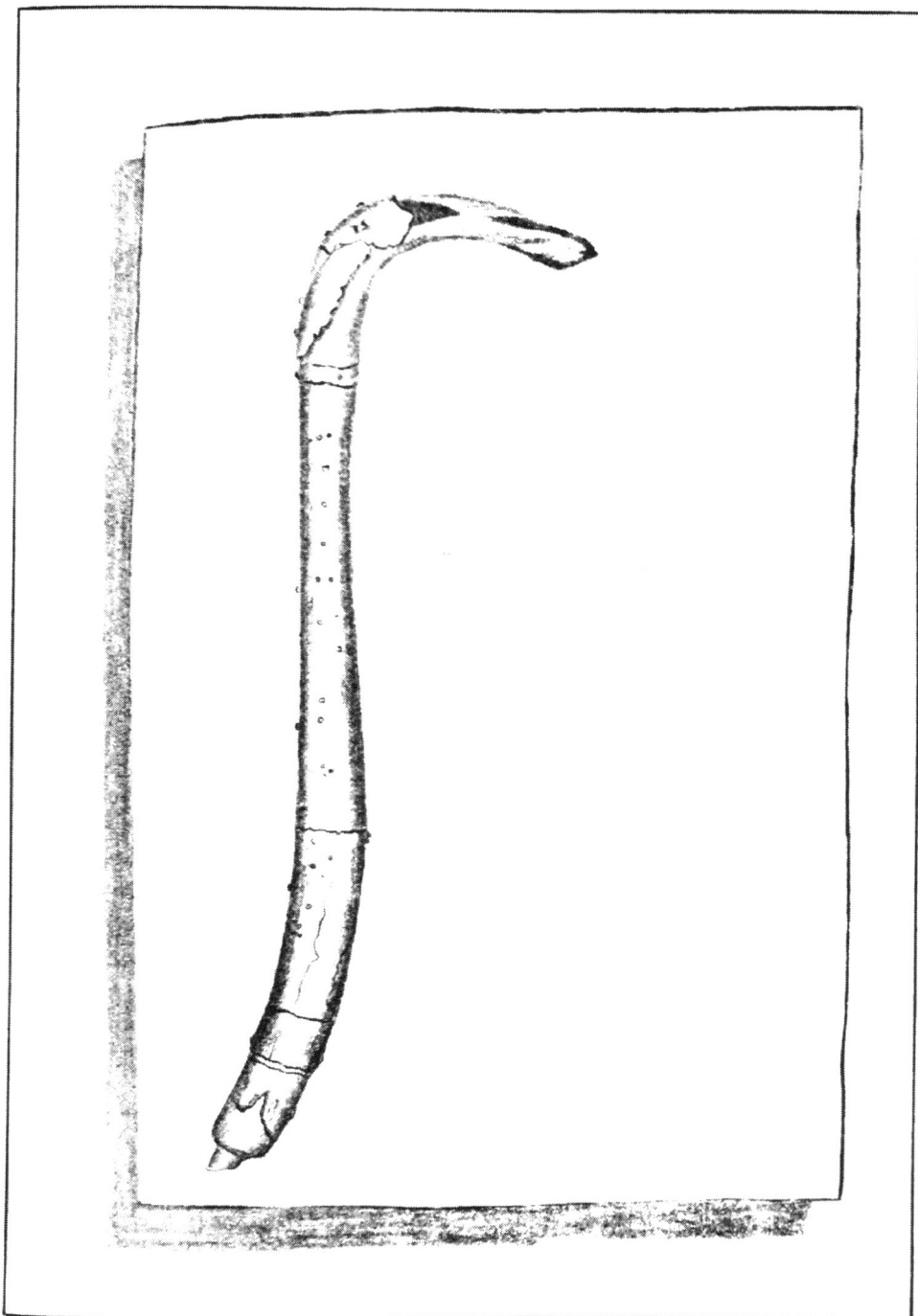

FOUND UNDER ROSEMARKIE RUNIC CROSS
(Supposed Crozier "or Bachial Mòr" of Moluag)

ANTIQUITIES.

FINDS IN FORTROSE AND ROSEMARKIE.

From a very remote period up to this very day, we have records of numerous finds being made at Fortrose and Rosemarkie; some of those finds have been valuable, others intensely interesting; each tending to unravel the unwritten history of this wonderful place of hoar antiquity. Almost daily, coins of gold, silver, and bronze are being picked up around the Cathedral precincts, and, were a thorough systematic search made, valuable results would undoubtedly accrue. One suggestion might be thrown out, viz., to have the foundations of the Cathedral better excavated to reveal the true internal construction of the building, and possibly valuable treasure trove might be discovered to compensate the trouble and expense.

Fortrose.

In the statistical account we find a record of 200 silver coins of the reign of Robert King of Scots being found in a copper jug of antique form, while digging the foundation of a house in the vicinity of the cathedral green; and in 1880 a brass ewer was found, containing 1100 silver coins of the reign of Robert III., King of Scots. Mr W. S. Geddie, Fortrose, reporting the discoveries to the Antiquarian Society, says :— "The places of coinage are Edinburgh and Perth, but at least one was marked ' *Villa : de : Aberd.*' " The diverse spelling in the inscriptions and other details show that they have been stamped with a variety of dies ; and among different readings of "Scotorum," in the legend round the head of the king, were "Scottorum," "Scotoru," and "Setrum"; also, "gratia," instead of "gra," &c.

Mr John Henderson, town clerk, has in his possession a silver coin of the reign of James I. It is about the size of half-a-crown, and was discovered by Mr Henderson's father some years ago. Mrs Henderson has a gold ring in her possession, found behind the post-office buildings. It is of the ordinary marriage ring type, quaintly carved in herring bone ornament, with an inscription on the inside which speaks for itself, viz., "*My : affection : is : my : affliction.*" Many small coins are frequently discovered, and numerous specimens can be viewed in the hands of several citizens. While some trenching was being done in the field known as the "Precincts," or site of the bishop's palace, foundation walls were discovered, covering an area of about one acre. The "Precincts" are advertised to feu as this is going to press; from its ancient historic memories, secluded genial situation, and fine old tree surroundings, we know of no more suitable place for the residence of an antiquary.

These and many minor discoveries have already been made at Fortrose, casually ; but, as previously suggested, were a scientific investigation instituted, more valuable results would necessarily follow.

Rosemarkie.

Perhaps the most interesting and valuable discovery we have recorded was made at Rosemarkie in the finding of the runic cross, or sculptured stone, which was found under the flooring of the parish church of Rosemarkie. The stone was appropriated by the contractor, and afterwards laid on his grave. It was substituted by a new grave-stone by a few interested antiquaries, and placed in its present position, at the north-west corner of the Parish Church. There is a story current that a crozier, or " bachiul mhor " (see plate), a ring, and crystal ball were found under the stone ; raising the supposition that Moluag was buried here, as St Columba gave such symbols of authority to this disciple only. The drawing facing this page represents the upper portion of both sides of the stone ; the two right hand panels belong to the west face, while the two on the left hand represent the east side. Below those panels intricate knotted work is seen, while all the edges exhibit similar ornamentation. It seems a pity to have so rare a specimen of Celtic art exposed to the defacing action of the elements—could it not find a spare corner inside the church, or a small ornamental covering made for it in the present position ?

In 1787 several silver coins were found in a small cairn of stones near Rosemarkie. They were mostly shillings of Queen Elizabeth, with a mixture of other coins, and particularly some of the time of James I. and Charles I.

A circular mound once existed on the terrace above the town, towards the west end ; but was considerably defaced, if not entirely destroyed, some years ago by erecting houses thereon. It was called Court Hill, and no doubt was the court where justice was dispensed in olden times.

The Rev. James Macdowall, parish minister, informs us that the early session records of the parish prior to 1730 were destroyed. There are two interesting silver communion cups belonging to the parish, given by a Countess Isabella of Seaforth, for the church of Rosemarkie in 1686, bearing a Latin inscription.

The market cross stood at the west end of the High Street, near the middle of the road. It was knocked down about fifty years ago by a load of hay, and broken to pieces. No trace of the fragments can now be found. It consisted of an octagonal sandstone slab, with a cap on top. The base of the cross was supported or built in a square box of stones, five or six feet square, and about two feet high. This support was removed some years previous to the smashing of the cross to build a well dug close by.

ROSEMARKIE RUNIC CROSS.
(From a Drawing).

APPENDIX

ON

Cromarty and other Places of Interest in the Black Isle.

Cromarty.

In addition to the excursions described in the foregoing pages, the visitor might profitably make a trip to Cromarty, the objects of interest in the vicinity of which are here briefly described.

Cromarty, the birthplace of Hugh Miller, is situated at the extreme north end of the Black Isle, at the entrance to the firth which bears its name. It is about 8½ miles by road from Fortrose, but during the summer months the Black Isle steamers make occasional trips. On a fine day the sail down the firth is most enjoyable, and a fine view of the rugged cliffs bounding the coast can be obtained.

Cromarty, in early times, was a place of considerable importance (although now its ancient glory is faded and gone), and one of its proprietors, Sir Thomas Urquhart, will ever be known to English scholars by his inimitable translation of the 1st and 3d Books of Rabelais.* During the winter storms many ships

---

* "Sir Thomas was born about 1613, and died in 1660, the cause of his death being an immoderate fit of laughter, combined with the effects of 'flowing cups.' As is evident to any who reads his translation of Gargautua and Pantagruel. Sir Thomas was a man of lively fancy, great learning and wit. He had a weakness for genealogy, however, and he traced his own family up to Adam, himself being the 153rd in descent! By the mother's side he connected the different links back to Eve! The first of the family that settled in Scotland was one Nomostor (married to a daughter of Alcibiades), who took his farewell of Greece and arrived at Cromarty, or *Portus Salutis*, 389 years B.C.! Sir Thomas was

run here for shelter, guided by a lighthouse (its fixed red light being visible thirteen nautical miles), built at a cost of £3203. Cromarty Roads are considered the safest on the British coast; blow as it may, the largest vessel can ride in safety within this natural harbour. According to history, so early as the sixteenth century it was "callit by the Scottish folks the haill of seamen," while to the ancients it was known as *Portus Salutis.* From the fact that the principal family surnames are peculiar to the south, it may be inferred it was originally colonised by southrons.

Prior to the opening of the railway to the North, it was the chief mercantile marine centre of the northern counties, but now its ancient greatness has dwindled down, and the busy whirr of machinery, or the noisy din of hurrying workmen, is heard no more, but all is quiet and silent as a deserted city; where once hundreds of people found daily employment now scarcely can the poor eke out a miserable existence. Hugh Miller says that no fewer than five three-masted vessels and a large number of lesser ones belonged to the port in his day. Whence then this decay? Simply isolation. Placed as it is at the extreme end of the peninsula, completely cut off from the outer world, need one wonder at its decline? Give Cromarty railway connection and it will revive, its fishing develop, and it may not be too much to expect that, with its fine harbour, it may become the chief maritime station for continental trade in the north of Scotland.

When the museum of the Northern Institution was in existence, there was an ancient custom house seal or cocket of the united burghs of *Invirnis et de Chrombhte,* supposed to belong to the reign of Robert II. Where the seal and the museum now are would be interesting to know, but it seems a pity that so

---

knighted by Charles I., and was present with Charles II. at the battle of Worcester (1651), where he was taken prisoner by the army of the Commonwealth, and where about 100 MSS., containing the results of his studious hours, were lost." A periodical critic of the beginning of this century describes Sir Thomas as "not only one of the most curious and whimsical, but the most powerful also, of all the geniuses our part of the island has produced." He was proprietor of nearly the entire shire of Cromarty, while now not an acre in the country is owned by any of his name or descendants.

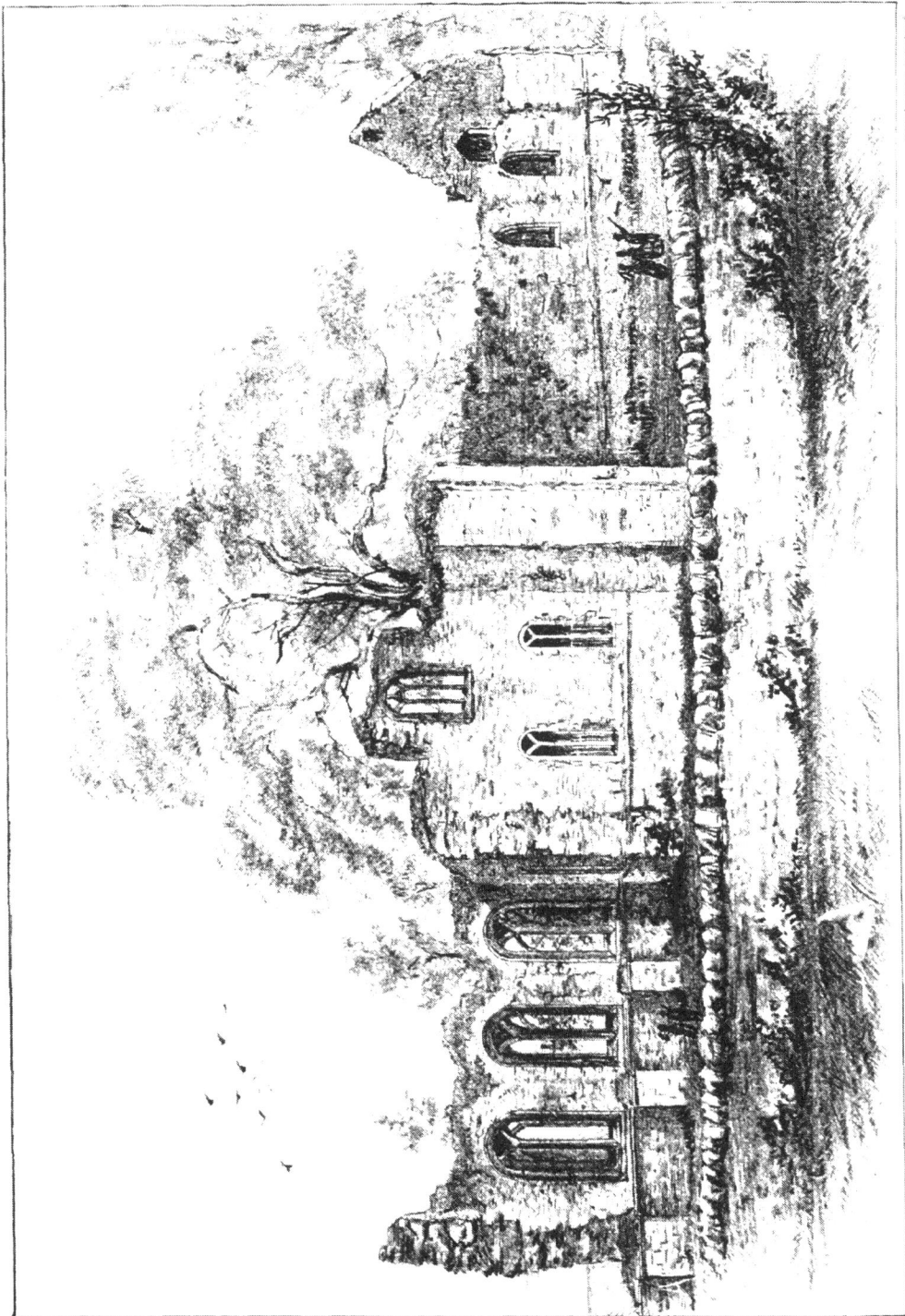

BEAULY PRIORY.

useful an institute as a museum for northern antiquities should have been allowed to collapse. On many occasions Scotch and French coins have been dug up in the gardens and fields about.

The other places of interest round and about Cromarty are: the ruins of Dunskaith Castle, on top of a precipitous rock on Northern Sutor, immediately opposite Cromarty; old graveyard; site of St Relugus' Chapel (the Trinitarian Priory, erected in 1271, has entirely disappeared); house in which Hugh Miller was born, and his monument (a column 40 feet high) surmounted by his life-size statue, by Handyside Ritchie. The present House of Cromarty is built on the site of the old castle of the Urquharts, which was six storeys high, stone-roofed and battlemented, razed to the ground in 1772. The Fiddler's Well, Gallow Hill, South Sutor, supposed to contain encampment of Sir William Wallace; St Bennet's Well, and ruin of Chapel, near Eathie Burn; Morial's Den, along shore towards Jemimaville (see Hugh Miller's "Schools and Schoolmasters"); several small earth forts on ridge, north side of road leading to Fortrose; Clach Malloch, or Cursed Stone, mentioned by Hugh Miller that within the lifetime of men then living they recollect steering the plough when young, where, when old, they steered with the rudder, the sea having encroached so much on *terra firma*.

Population in 1800 . . . 2208.
  „ 1881 . . . 1996.

Other Places of Interest in the Black Isle.

We here supplement a list of places of interest and antiquity in the Black Isle, not embodied in the foregoing pages. As many of them are well worthy of a visit, we cursorily describe several places, and, as they are noted on the map, a geographical description seems unnecessary, but, for the sake of continuity in reference, each parish is taken separately, beginning in the extreme west with the

Parish of Urray.

Convenient Centre—Muir of Ord Village or Beauly.

Beauly (*Beaulieu*, Fr. " beautiful place "), although situated in the county of Inverness and parish of Kilmorack, may be classed within our province, and in alluding to the antiquities of the Black Isle, it might seem a gross omission in not referring to the Beauly Priory.

> " I do love these ancient ruins—
> We never tread upon them but we set
> Our feet upon some reverend history ;
> And, questionless, here some men lie interr'd,
> Loved the Church so well, and gave so largely to it,
> They thought it should have canopied their bones
> Till doomsday ; but all things have their end—
> Churches and cities, which have diseases like to men,
> Must have like death which we have."

The village, which is fast aspiring to a more dignified title, is neat, clean, and pleasantly situated, and a convenient centre for making tours to Strathglass, Glencannich, Strathconan, and the mountain ranges of the shires of Inverness and Ross. Situated at the east end is the ruined priory of St John Baptist. It was founded in 1232, by Sir John Bisset of Lovat, for seven monks, a sub-order of the Cistercians, who followed the rule of St Benedict. Its aisleless church is 136 feet by 21 feet, of the early 2d Pointed, and may date from the first decade of the 14th century. The last Prior granted its lands in 1558 to the sixth Lord Lovat ; but, forfeited by Alexander Mackenzie of Fraserdale in 1716, they are now Crown property. There are many superstitious tales told of this, as is the case with most ruined buildings, the most weird story being of the tailor who ventured to sew a pair of trousers at midnight inside the building, and barely escaped from " Auld Nick," who left the imprint of his hand on the door, the effect of a missed slash at the *tail-leur* as he bolted through the doorway. E. Chisholm-Batten's " Beauly Priory " gives a most interesting and valuable account of the building. On the hill behind the village are the remains of a circular encampment, called *Dun-mor*, where Montrose is said to have rested his army.

Muir of Ord village, called Tarradale, is a modern village of considerable size, sprung up since the opening of the railway. It has two good hotels. Large fairs are held monthly near the village, and during summer the Inverness Militia encamp for training near the market stance.

Ancient Earth Fort, near Muir of Ord Station; interior of, 84 by 64 feet, with surrounding ditch or trench 20 feet wide.

Clach-an-Seasaidh, or the Standing Stones, at south end of Muir of Ord Market Stance; associated with Coinneach Odhar, the Brahan Seer's prophecy, "That the raven will drink, for three successive days, the blood of the Mackenzies off the top of the stones."

Cille Chriosd.—Chapel restored, in which the Macdonells of Glengarry burnt the Mackenzies by setting fire to the chapel one Sunday morning during worship. The tragedy is known as the "Raid of Gilchrist," see Anderson's "Guide to the Highlands;" the other antiquities being Fairburn Tower, Stone Circle at Arcan, Kinkell Castle, erected in 1614, Site of Ancient Chapel at Orrin Bridge. "David's Fort," situated in wood above Conan House, and below Bishop Kinkell, is an earth structure of rhomboidal form, with outer wall, surrounding ditch, and inner fort or rampart, supposed to be a Roman construction; size of inside 80 feet, average depth of ditch 9 feet to 12 feet by 15 feet wide (see proceedings of Antiquarian Society, 1883.)

Mansions—Fairburn House (modern), seat of Mr Stirling.

Ord House, seat of Mr Mackenzie.

Highfield, seat of Mr G. F. Gillanders.

Tarradale House, birthplace of Sir R. Murchison.

Population in 1800 2083

      "   1881 2393

of whom 1681 were Gaelic speakers.

Killearnan Parish.

Convenient Centre—Milltown Village.

Milltown village is situated on the Beauly Firth, near Redcastle; has some nice houses, public school, and two churches in vicinity.

ANTIQUITIES.

Carn Inenan, Stone Circle, on Millbuie ridge, consisting of three rings, outside one 66 feet diameter. *Cairnglas*—huge cairn of stones, said to mark the grave of a Danish prince. Numerous cairns and tumuli in vicinity. Stone coffin and skeleton discovered near Cairnglas, 25th August 1881.

Bee-hive Dwellings and burial cairn near Croftcrunie, Drumnamarg. "The Temple," (*vide* "Proceedings of the Society of Antiquarians of Scotland," June 1882, 484-88).

Kilcoy Castle.—A very good specimen of the castles of the 16th and 17th centuries; "a simple quadrangular structure, with overhanging turrets at angles, and with round towers at the diagonal corners, by which the side walls are effectually flanked and protected." The small entrance door is also strongly guarded. The roof of ground floor is vaulted in stone, while the lintel over fire-place in dining room is elaborately carved with three coats of arms, and bears the date 1679 (*vide* "Inverness Scientific Society Transactions.")

Redcastle.—Said to be the oldest inhabited house in Scotland, its ancient name being Eddyrdor, was erected in 1179 by William the Lion, who along with his brother David went to Ross with a large army to quell some disorder in that distant quarter, and built two castles, and fortified them. "In 1481 James III. granted to his second son, James Stewart, Marquis of Ormond, the lands of Avauch and Netherdal, with the moothill of Ormond, and the castle and fortalice of Redcastle." In 1278 it was in possession of Sir Andrew de Boscho and his wife Elizabeth, and they two paid two merks yearly to the monks of Beauly at their castle of Eddirdovar. It afterwards passed into the hands of Sir John Bisset, the founder of Beauly Priory. It is now considerably modernised, yet it presents a massive and dignified appearance, although in detail it does not quite equal Kilcoy Castle. Redcastle is the seat of the Hon. Henry Baillie. Opposite Redcastle, and a considerable distance below high water mark, are several cairns, in some of which human bones were discovered. The parish is the birthplace of General Sir George Elder, General Mackenzie Fraser, and Sir Roderick Murchison, the father of geology. Sandstone quarry at Milltown village, from which stone used in construction of Caledonian Canal was procured.

Population in 1800 1131
    ,,   1881 1059
of whom 558 were Gaelic speakers.

Kilmuir Wester & Suddie.

Convenient Centre—Munlochy.

Munlochy.—This pretty little village, now fast rising into importance, is pleasantly situated at the head of Munlochy Bay, nestling picturesquely among the variegated foliage of the chestnut, the oak, and the elm. There are numerous pleasant and invigorating walks about, and the scenery presents the appearance of an English or lowland landscape, for nature, with

> " Her patient hand, with ever pleasing change,
> Adorns the cultured walks of Allangrange,
> Or spreads along Redcastle's fair domain,
> Each varied charm of water, wood, and plain."

The village has improved very much recently, being erected into a special water and drainage district. The supply of water is plentiful, and the sanitation good. Munlochy is distant five miles from Kessock Ferry, and five from Fortrose. A mail coach runs daily, so that to any one wishing a quiet rural retreat this locality can be recommended. The village is well-lighted, has a public library and reading-room, branch of the Royal Bank, post, telegraph, and money order office, a spacious hall, and a corps of volunteers. A new Free Church is being erected, which will be a very ornamental edifice. The other villages in the parish are, Kilmuir, now resorted to as a summer retreat, and Kessock village and Charleston, both in the vicinity of Inverness. Highland games are held annually at Munlochy in August, and the Black Isle Volunteer Wapinschaw takes place here in June.

ANTIQUITIES.

Stone Circle in wood near Muirtown ;* Stone Circle at Braedoune, near Belmaduthy; remains of Knights Templars' Chapel at Allangrange Old House ; remains of Old Churches in the grave-yards of Suddie and Kilmuir ; Hut Circles at Teandore, head of the Allangrange valley *(vide* Transactions Soc. Antiq. Scot.)

* Captain C. M. Cameron, Balnakyle, informs us that the piece of land occupied by this stone circle is always reserved when leasing the farm on which it stands.

James' Temple—Caistail Shemus—remains of circular structure on summit of Drumderfit Hill. All along the summit of this ridge are numerous remains of very fine hut circles, mounds, and tumuli, with peculiar parallel ridges or dykes connecting the huts.

It may be interesting to note that Donald of the Isles once encamped in this parish with a strong force, intending to attack and burn Inverness, and accordingly sent notice to the provost to that effect. That dignitary hastened with some attendants to offer Black Donald conciliatory terms. Meanwhile the provost's followers managed to smuggle and liberally distribute whisky into camp, the result being an easy victory for the Invernessians; every man was slaughtered, except one fellow, who hid beneath a *Carn Lobban*, a sort of basket cart without wheels, then greatly used in the Highlands. He was afterwards known as Lobban, latterly changed to Logan. This man settled down there, and we find it chronicled that Drumderfit was occupied in 1564 by one William Lobban, while it is a prevalent maxim in the district " As old as the Lobbans of Drumderfit."

On the furthest west shoulder of Drumderfit Hill is a large cairn of stones called *Blair-na-coi*, or the battlefield of the *coi*—the cross beam or yoke coupling oxen—from the circumstances that the man who " turned the battle " belaboured his opponents with this beam, having hurried from his work at the plough on seeing the battle go against his countrymen. The battle is said to have occurred in 1340.

Vitrified Fort on Ord Hill, immediately above Kessock Ferry, and *Earth Fort* near Loch Lundie (see page 9).

According to ancient records, a chapel existed in this parish at " Haldach," or " Handach," also called " Haldact, with the Inn of the same, callit Tolgormok." Strange that no trace of this place can be found—not even a locality of this name is known, although we are of opinion that Torgorm Point, west of Charleston, yet outside the present parish boundary, can be identified with the site. The grounds for this supposition are, that portions of Kilmuir Wester and Killearnan parishes were excambated, and about two years ago human remains were found at this place, while no other name in this or adjoining parishes bears so much resemblance to Tolgormok as Torgorm Point.

The present parish church was built in 1754, but since renewed. The two old churches of Kilmuir and Suddie are now in ruins, their yards being used as the parish burying grounds. There is a story current that the old bell from Suddie Church was stolen one night, but the jaggling of the boat rung the bell so violently that the men were forced to throw it overboard somewhere near Craigiehow.

Geology.—There is nothing remarkable about the geology of this parish, except perhaps the distinct traces of ancient raised beaches, with deposits of shell from seventy to ninety feet above sea level, and the clay deposits of the Munlochy and Allangrange valleys. The prevalent rock is old red sandstone and conglomerate. The very fine sandstone quarry of Suddie is noticed by the late eminent geologist, Dr Page, as possessing the finest building stones in the north of Scotland.

General John Randall Mackenzie, who fell at the battle of Talavera in 1809, was a native of the district; and Sir William Fairbairn, C.E., received his early education at Munlochy school, his father residing at Allangrange.

Mansion Houses.—Belmaduthy, Allangrange, and Drynie.

Population in 1800 1703
„ 1881 1866
of whom 1071 were Gaelic speakers.

Parish of Avoch.

Convenient Centre—Avoch or Fortrose.

The following places are described in Excursion No. I.; they are merely classified here:—

Ormond Castle; Bennet's Monument; Quarry and Craigack Well; Rosehaugh House; Avoch House; Arkendeith Tower; Bishop's Well; Petrifying Stream; Fossil Bed; Ancient Grave Yard; Rare Plant *(Pinguicula Alpina).*

Population in 1800 1476
„ 1881 1691

Parishes of Rosemarkie & Cromarty.

Fully noticed under their respective heads.

We will now travel westward, along the north side of the peninsula, through the parish of

Kirkmichael & Cullicudden.

Among the antiquities of which we find the ruins of Kirkmichael Chapel, and remains of chapel at Balblair. Numerous mounds and tumuli are scattered all over the parish.

In one of those broken into at Jemimaville, an earthen urn was discovered; while in trenching some land on the glebe a stone cup, 4 inches in diameter and ¾ of an inch thick, was found embedded about 18 inches below the surface, inside the circular remains of a Pictish house.

Kinbeachie Castle—in ruins.—South of Kinbeachie, and between it and the watershed, are numerous tumuli scattered over the moor, some very large.

The Ruins of Castle Craig.—Built on a rock close to the Cromarty Firth; there are traces of a surrounding wall; only one wing of the building is now standing, and all the rooms were vaulted in stone. There is little more than traditionary history connected with the castle, but we find in the statistical account of Scotland "that the building was originally erected by the Urquharts, barons of Cromarty; and one of this family, by his misconduct falling under the censure of the Church, is said to have been deprived of the castle and lands in the immediate vicinity thereof, and which eventually became the property of the Church." That this castle was the principal residence of the Bishop of Ross is quite certain. "An ancient document is now (1836) in the Museum of the Antiquarian Society at Inverness, presented by Colin Mackenzie, Esq. of Newhall. It is a warrant signed by the Bishop of Ross, and dated at Craighouse, his residence; in virtue of which certain persons were to be pursued and incarcerated for violently resisting the possession of the place of Tolly, near Dingwall, to whom the bishop granted a lease of it." The castle eventually passed into the hands of the Williamsons; and afterwards possessed by the Roses of Kilravock, an ancient family, then possessing much property throughout the Black Isle.

Population in 1800   1200
     ,,   1881   1424

of whom 601 are Gaelic speakers.

Mansions.—Newhall, Poyntzfield, St Martins, and Braelangwell.

Urquhart and Logie Wester.

Ferrintosh.—The district was originally comprehended in the extensive county of Inverness; it still pays cess or land tax in that county. In the 15th century it was the property of the Thane of Cawdor, who procured an annexation of that, with other lands, to the county of Nairn. Accordingly, when the shire of Ross was erected, we have seen that this district was exempted from it. The estate (which now belongs to Mr Forbes of Culloden) was at one time famous for enjoying the privilege of distilling barley of its own growth into spirits, free of duty; and although the right was purchased by the Government about ninety years ago (1790) the skill of the art still remains among the people.

"The lands belonging to Mr Forbes of Culloden, which go by the name of Ferrintosh, and form the central and largest division of the parish, possessed, from 1690 to 1786, an exemption from the duties of excise on spirits distilled from grain of their own growth. This privilege was originally granted to Duncan Forbes, one of the patriots who, at the glorious period of the revolution, stood up in defence of the religion and liberties of their country. By opposing the disaffected, and supporting the loyal subjects in his neighbourhood at much expense, he was materially instrumental in quashing the rebellion which at that time threatened the north of Scotland.—" New Statistical Account of Ross and Cromarty."

Among the antiquities are the following, viz.:—Cairn in wood above Culbokie, adjoining Culbokie Loch, in which human remains were found in 1848. Cairn near Glascairn Farm. Circular Fort in wood above the famous preaching stance in Ferrintosh burn. Huge tumuli near the Free Church School and Manse. According to the "Origines Parochiales Scoticæ," several conical cairns in the south-west of Logie parish—opened about 1795— three stone coffins, ranging in a line from east to west, were found. According to the same account, St Malrube of Apple-cross is said to have been murdered in 722 by the Norwegians at Urquhart, in Ross. There was erected, says the "Aber-deen Breviary," on the spot where he was slain, a chapel of oak, which afterwards became the parish church of Urquhart. The ruins of the ancient church is still in fair preservation, and is in the county of Ross, while the present parish church is situated in a detached part of the county of Nairn.

Although this parish within itself offers little in the way of scenery, still a magnificent panorama of mountain, glen, and sea may be obtained from the summit of its slopes—

> " In rugged grandeur by the placid lake,
> Rise the bold mountain cliffs, sublimely rude ;
> A pleasing contrast, each with each, they make ;
> And, when in such harmonious union viewed,
> Each with more powerful charms appear imbued.
> Ever thus it is, methinks, with mingling hearts ;
> Though different far in nature and in mood,
> A blessed influence each to each imparts,
> Which softens and subdues, yet weakens not, nor thwarts."

while its old associations, superstitions, lore, and antiquities will amply supply the deficiency which nature's adorning hand seems to have passed unnoticed. In no parish in Scotland is local superstitions and folklore so current, and not a bleak moor-land, dingy dell, or yawning gorge is without its *phantom, baobh,* or *brownie,* while the Ferrintosh witches are still credited with playing at the itinerant game of scouring the peninsula in the shape of hares.

Population in 1800 4430
„ 1881 , , , , 2083

DIRECTORY

OF

FORTROSE & ROSEMARKIE.

MAGISTRATES, TOWN COUNCIL, &c.—Provost, Peter Grant; Bailies, Alexander Hossack, G. Sutherland, and Alexander Watson; Dean of Guild, J. Hossack; Treasurer, Wm. S. Geddie; Councillors, A. Fraser, A. Henderson, M. Home, A. Maciver, J. Smith, A. Mackenzie, Andrew Grant, G. Paterson, and John Gordon; Town Clerk, John Henderson; Chamberlain, R. J. Gillanders; burgh officers, George Ross and John Gordon; burgh meter, Capt. W. Grieve; tacksman of petty customs, Alexander Watson; assessor of lands and heritages, F. Foster, Inverness.

POLICE COMMISSIONERS.—The burgh has adopted the Lindsay Police Act, and the Magistrates and Town Council are the Commissioners under the Act. Clerk to the Commissioners, John Henderson; sanitary inspector, George Ross; water inspector, Joseph Allan.

SHERIFF CIRCUIT SMALL DEBT COURT.—Circuit Small Debt Courts are held quarterly by the Sheriff-Substitute of Dingwall. Sheriff-Clerk Depute, John Smith, Fortrose.

JUSTICE OF PEACE COURT.—The neighbouring Justices meet for the trial of causes within their jurisdiction on the first Wednesday of each month. Justices in Fortrose and neighbourhood:—General Macintyre, Fortrose; A. R. Mackenzie, M.D., Fortrose; the Provost, Senior Bailie, and Third Bailie; Rev. J. Gibson, Avoch; J. Fletcher, Esq. of Rosehaugh; R. G. Mackenzie, Esq. of Flowerburn; depute clerk, J. Henderson.

PLACES OF WORSHIP.—Church of Scotland, Rosemarkie, Rev. James Macdowall; *Quoad Sacra* Church of Scotland, Fortrose, Rev. R.

O. Young. Free Church, Fortrose, Rev. Charles Falconer. St Andrew's Episcopal Church, Rev. Spence Ross. Baptist Chapel, Rev. Ferdinand Dunn. Services—forenoon and afternoon.

SCHOOLS.—Fortrose Academy, Charles Laverie, rector ; Rosemarkie Public School, David Cunningham, M.A., teacher ; Infant School, Fortrose, Mrs Catherine Jeffrey, teacher.

PAROCHIAL BOARD.—Chairman, Major Mackenzie of Flowerburn ; inspector of poor and collector of rates, R. J. Gillanders ; medical officer, Dr Mackenzie, Fortrose.

SCHOOL BOARD.—J. Douglas Fletcher, Esq. of Eathie (chairman); Dr Mackenzie, Rev. J. Macdowall, Bailie Alexander Hossack, Fortrose ; and Kenneth Mackenzie, Courthill, Rosemarkie ; clerk and treasurer, J. Henderson ; officer, Alexander Macallan.

HARBOUR TRUSTEES.—The Magistrates and Town Council act as Harbour Trustees under the Fortrose Pier and Harbour Order (1879). Harbour and shore master, Capt. Grieve.

BLACK ISLE STEAM SHIPPING CO., LIMITED.—Directors, J. Douglas Fletcher, Esq. of Rosehaugh (chairman); R. G. Mackenzie of Flowerburn ; Bailie Hossack, Fortrose ; John Smith, Fortrose ; A. Strother, Inverness ; Chas. Innes, do ; Major George Rose, do. J. Henderson, Fortrose, manager and secretary. A steamer is run to Inverness three times a week during the winter and spring, and daily during the summer and autumn months (see advertisement).

THE ACADEMY.—Resident directors are, James Fletcher, Esq. of Rosehaugh ; and R. G. Mackenzie of Flowerburn. Donors of 50 guineas are directors, and their male heirs in succession. Donors of 20 guineas are visitors for life. Clerk and Treasurer, J. Henderson. A scheme under the Educational Endowments (Scotland) Act, 1882, has been drafted for placing this institution under the School Board of Rosemarkie as a higher class school, which will be shortly carried into effect.

REGATTA.—Commodore, R. G. Mackenzie, Esq. of Flowerburn ; treasurer, Provost Grant ; joint secretaries, A. Mackenzie, Esq. of Breda, and Thomas Henderson, Esq., Suddie. A regatta is held in the Fortrose bay every year, about the beginning of August (this year—1885—to be held on the 15th August).

POST OFFICE.—Postmaster, John Smith ; receiving office, Rosemarkie, Kenneth Mackenzie ; receiving office, Avoch, F. Finlayson ;

contractor for the mails between Inverness and Fortrose, J. Munro, Fortrose. The mail gig leaves Fortrose at 7 A.M., and arrives at Kessock at 8.15 A.M.; leaves Kessock at 1.30 P.M., arrives at Fortrose at 2.45 P.M.

BLACK ISLE COMBINATION POOR-HOUSE.—Parishes in the Union—Rosemarkie, Cromarty, Avoch, Knockbain, Resolis, Killearnan, and Urquhart. The Managing House Committee consist of two members from each Board. House Committee—Rosemarkie; R. G. Mackenzie of Flowerburn, Rev. J. Macdowall. Cromarty, D. M. Ross of Cromarty, and J. Middleton. Avoch, J. Fletcher of Rosehaugh, and Rev. John Gibson, Avoch. Knockbain, J. Mackenzie of Allangrange, and Rev. J. Macgregor of Knockbain. Resolis, Rev. R. Macdougall, and C. Lyon Mackenzie. Killearnan, Capt. C. M. Cameron, Balnakyle, and R. Trotter, Garguston. Urquhart, D. Forbes of Culloden, and Rev. M. Macgregor, Ferrintosh. Chairman of Committee, James Fletcher of Rosehaugh; vice-chairman, Rev. J. Gibson; governor, John Macdonald; matron, Mrs Macdonald; medical officer, A. R. Mackenzie, M.D.; secretary and auditor, Robert J. Gillanders.

BLACK ISLE FARMERS' SOCIETY.—Patron, James Fletcher, Esq. of Rosehaugh; president, C. Lyon Mackenzie; vice-president, A. T. Macqueen, Coulmore; acting committee, J. D. Fletcher, William Murray, C. M. Cameron, F. Lawson, D. M. Ross, and H. Maciver. Meet at Fortrose and Munlochy. Hon. secretary and treasurer, Thomas Henderson, Suddie.

ASSESSED TAXES COMMISSIONERS FOR THE BLACK ISLE DISTRICT.—The Commissioners of Supply within the district.

CALEDONIAN BANK—Peter Grant, agent; Thomas Murray, Accountant.

MECHANICS' INSTITUTION, READING-ROOM, & LIBRARY. Patron, James Fletcher, Esq. of Rosehaugh; president, Rev. James Macdowall, Rosemarkie; vice-president, J. Henderson; secretary, T. Murray; treasurer, J. M. Stuart; librarian, F. Tout.

ROSEMARKIE MUTUAL IMPROVEMENT SOCIETY.—President, Rev. J. Macdowall; secretary, Mr D. Cunningham; treasurer, Mr Hugh Thomson, Fortrose.

CHANONRY LODGE OF GOOD TEMPLARS—Fortrose, No. 628. Office-bearers, M. Home, W.C.T.; R. Home, W.V.T.; R. Elder, W.S.;

Alex. Macallan, W. Chap.; Donald Sinclair, W.T.; James Turnbull, W.M.; A. Rennie, W.I.G.; A. Fraser, L.D.

ROSEMARKIE COAL AND CLOTHING CLUB.—James Fletcher, Esq. of Rosehaugh, patron; Mrs Grant, secretary; Dr A. R. Mackenzie, treasurer.

AUXILIARY BIBLE SOCIETY.—James Fletcher, Esq. of Rosehaugh, patron; P. Grant, Esq., treasurer.

BLACK ISLE HORTICULTURAL SOCIETY.—President, J. Douglas Fletcher, Rosehaugh; vice-president, Major Smith, Munlochy; convener of committees, Rev. S. Ross, Avoch; secretary, J. S. Glass, Avoch; treasurers, W. S. Geddie and T. Murray, Fortrose.

Population of burgh, 986. No. of voters, 179. Valuation £3603 16s. 9d.

ACCOMMODATIONS.—Fortrose, Royal Hotel, Andrew Grant proprietor; Temperance Hotel, C. Junor, proprietor. Rosemarkie, Brae House, Mrs Forsyth; Ivy House, J. Hossack; Poplar Villa, Miss Gillanders; Court Hill Cottage, K. Mackenzie; Shore Cottage, Mrs Junor; Miss M. Home, High Street; D. Dunoon, High Street; William Home, High Street; Mrs Mackenzie, do.; Mrs H. Wilson, do. In addition to above the following residences are let as summer quarters, most of them being suitable for families—Glenmutchkin House, beautifully situated up the Fairy Glen; Hawkhill House, Crawmarkie, Pebble Villa, Seabank Cottage, Marine Cottage, Rose Cottage, Thistle Cottage, Craigbanks Cottage, Malord Cottage, Flowerburn Cottage, Meadowbank.

THE END.

BLACK ISLE STEAM SHIPPING

COMPANY (LIMITED).

THE swift Screw Steamer, "ROSEHAUGH," plies daily between Fortrose and Inverness from May till October inclusive, and bi-weekly during the other months of the year. For the convenience of Tourists, the " Rosehaugh " makes

SPECIAL TRIPS

during the Season, on MONDAYS, THURSDAYS and SATURDAYS. The Steamer has ample Cabin and Steerage accommodation, and the Fares are moderate.

For further particulars see Advertisements and Time-Bills.

JOHN HENDERSON,

MANAGER, FORTROSE,

DRILL HALL,
MUNLOCHY.

Hall 60 ft. by 30 ft., with spacious Ante-Room and
Moveable Platform.

Will be Let for

CONCERTS, BALLS, LECTURES, AND OTHER MEETINGS

on Moderate Terms.

SCALE OF CHARGES and all information on application to
LIEUT. ROBERT GREIG, Munlochy.

W. S. GEDDIE,
DRAPER, GROCER, AND GENERAL MERCHANT.
HATS AND CAPS,   BOOTS AND SHOES, STATIONERY.

Particular attention paid to the DRESSMAKING, MILLINERY,
and MANTLEMAKING.

EXCHANGE BUILDINGS,
THE CROSS, FORTROSE.

JOHN AVERY & CO.,

LIMITED,

Printers, Lithographers, Engravers,

Die-Stampers, Bookbinders,

Wholesale Stationers, Publishers,

&c., &c.

ABERDEEN.

Estimates given and Designs Furnished on Application.

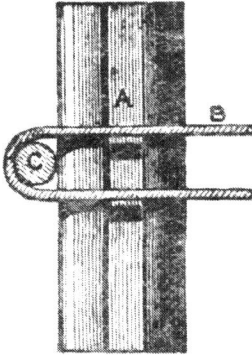